Cell Cycle Regulation

NEW HORIZONS IN THERAPEUTICS
SmithKline Beecham Pharmaceuticals U.S. Research Symposia Series

Series Editors: Robert R. Ruffolo Jr, George Poste and Brian W. Metcalf
SmithKline Beecham Pharmaceuticals, Philadelphia, Pennsylvania

---

CELL CYCLE REGULATION
Edited by Robert R. Ruffolo Jr, George Poste and Brian W. Metcalf

This book is part of a series. The publisher will accept continuation orders which may be cancelled at any time and which provide for automatic billing and shipping of each title in the series upon publication. Please write for details.

Cell Cycle Regulation

Edited by
ROBERT R. RUFFOLO Jr
GEORGE POSTE *and*
BRIAN W. METCALF
SmithKline Beecham Pharmaceuticals
Philadelphia, Pennsylvania

harwood academic publishers
Australia • Canada • China • France • Germany • India • Japan
Luxembourg • Malaysia • The Netherlands • Russia • Singapore
Switzerland • Thailand • United Kingdom

Amsteldijk 166
1st Floor
1079 LH Amsterdam
The Netherlands

British Library Cataloguing in Publication Data

Cell cycle regulation. - (New horizons in therapeutics ; 1)
  1.Cell cycle 2.Cellular control mechanisms
  I.Ruffolo, Robert R. II.Poste, George III.Metcalf, Brian W.
  571.6

ISBN 9057022044

Contents

Chapter 1

Cell Cycle Regulation in Hematopoietic Stem Cells

Peter M. Lansdorp

Chapter 2

Coupled Control of Cell Proliferation and Cell Viability by Oncogenes and Cytokines

Gerard I. Evan, Nicola McCarthy, Moira Whyte, Lamorna Brown, Andrea Kauffmann-Zeh and Eugen Ulrich

Chapter 3

Angiogenesis and the Endothelial Cell Cycle

Jeffrey R. Jackson and James D. Winkler

Chapter 4

Complex Regulation of the NF-κB Transcription Factor Complex:
NF-κB Activation is Inhibitable by Tyrosine Kinase Inhibitors

Steven M. Taffet and Dalton Foster

Chapter 5

*Regulation of Inflammatory Cytokine Biosynthesis: Discovery of a
Low Molecular Weight Inhibitor and its Molecular Target*

John C. Lee, Sanjay Kumar and Peter R. Young

Chapter 6

Signaling by the Cytokine Receptor Superfamily

James N. Ihle

Chapter 7

Cytokine Driven Signal Transmission

*Gerald A. Evans, Roy J. Duhe, O.M. Zack Howard, Robert A. Kirken,
Luis Da Silva, Rebecca Erwin, Maria G. Malabarba and William L. Farrar*

Chapter 8

*The pp70S6k Signaling Pathway is Activated by a Complex Multi-step
Process Involving Several Signaling Pathways*

Timothy C. Grammer and John Blenis

Chapter 9

Biochemical Dissection of Nuclear Events in Apoptosis

Atsushi Takahashi, Emad S. Alnemri, Teresa Fernandes-Alnemri,
Yuri A. Lazebnik, Robert D. Moir, Robert D. Goldman, Guy G. Poirier,
Scott H. Kaufmann and William C. Earnshaw

Chapter 10

Bcl-2 Family Proteins in Cancer: Regulators of Cell Death Involved in Resistance to Therapy

John C. Reed

Preface to the Series

The unprecedented scope and pace of discovery in modern biology and clinical medicine present remarkable opportunities for the development of new therapeutic modalities, many of which would have been unimaginable even a few years ago. This situation reflects the unparalleled progress being made not only in the disciplines of pharmacology, physiology, organic chemistry, and biochemistry that have traditionally made important contributions to drug discovery, but also in newer disciplines such as molecular biology, molecular genetics, genetic medicine, cell biology, and molecular immunology that are now of sufficient maturity to further our understanding of the pathogenesis of disease and the development of novel therapies. Contemporary biomedical research, embracing the entire spectrum of biological organization from the molecular level to whole body function, is on the threshold of an era in which biological processes, including disease, can be analyzed in increasingly precise and mechanistic terms. The transformation of biology from a largely descriptive, phenomenological discipline to one in which the regulatory principles underlying biological organization can be understood and manipulated with ever-increasing predictability, brings an entirely new dimension to the study of disease and the search for effective therapeutic modalities. In undergoing this transformation into an increasingly mechanistic discipline, biology and medicine are following the course already charted by the sister disciplines of chemistry and physics.

The consequences of these changes for biomedical research are profound and have resulted in new concepts, new and increasingly powerful analytical techniques; new advances generated at a seemingly ever-rapid pace; an almost unmanageable glut of information dispersed in an increasing number of books and journals, and the task of integrating this information into a realistic experimental framework. Nowhere is the challenge more pronounced than in the pharmaceutical industry. Drug discovery and development have always required the successful coordination of multiple scientific disciplines. The need to assimilate many additional disciplines within the drug discovery process, the extraordinary pace of discovery in all disciplines, and the growing scientific and organizational complexity of coordinating increasingly ultra-specialized and resource-intensive scientific skills in an ever-enlarging framework of collaborative research activities represent formidable challenges for the pharmaceutical industry. These demands are balanced, however, by the excitement and the scale of the potential opportunities for achieving dramatic improvements in health care and the quality of human life over the next twenty years through the development of novel therapeutic modalities for effective treatment of major human and animal diseases.

It is against this background of change and opportunity that this symposium series, *New Horizons in Therapeutics*, was conceived as a forum for providing critical and up-to-date surveys of important topics in biomedical research in which significant advances are occurring and which offer new approaches to the therapy of disease. Each volume contains authoritative and topical articles written by investigators who have contributed significantly to their respective research fields. While individual articles discuss specialized topics, all papers in a single volume are related to a common theme. The level of these articles is advanced, and is directed primarily to the needs of the active research investigator and graduate students.

Editorial policy is to impose as few restrictions as possible on the contributors. This is appropriate, since each volume is limited to the papers presented at the symposium, and no attempt will be made to create a definitive monograph dealing with all aspects of the selected subject. Although each symposium volume provides a survey of recent research accomplishments, emphasis is also given to the examination of controversial and conflicting issues, to the presentation of new ideas and hypotheses, to the identification of important unsolved questions, and to future directions and possible approaches by which such questions might be answered.

The range of topics for the volumes in the symposium series is broad, and embraces the full repertoire of scientific disciplines that contribute to modern drug discovery and development. We therefore intend to publish future volumes in what we hope is viewed as a worthy series of compilations that reflects the excitement and challenge of contemporary biomedical research in defining new horizons in therapeutics.

Robert R. Ruffolo Jr
George Poste
Brian W. Metcalf

Preface to the Volume

This volume focuses on recent developments, and most importantly on the key discoveries, related to the cell cycle and its regulation, which we view as a critical new horizon in therapeutics. The papers that comprise this volume were originally presented at the Eighth SmithKline Beecham Pharmaceuticals United States Research Symposium held in King of Prussia, Pennsylvania in November 1995. The importance of this area of research has been understood for many years, inasmuch as entry into, or exiting from, the cell cycle are key components of a variety of homeostatic processes and pathological states.

Research into all aspects of cell cycle regulation has undergone explosive growth during the past decade, primarily due to the influence of the powerful techniques of molecular biology. We are now beginning to understand the myriad of enzymes that are responsible for regulation of the cell cycle, as well as the temporal relationships of their actions in the various stages of the cell cycle. Indeed, many components of the cell cycle have been cloned and characterized in detail, and new components of the cell cycle are being identified on a seemingly daily basis. At the same time, there has been a quantum leap in our understanding of the genes that are activated or suppressed during various stages of the cell cycle, as well as the cellular processes that determine whether the cell cycle is activated or switched off. Most importantly, there has been significant progress made in identifying drugs that act on different components of the cell cycle, and as a result, medicine may soon have the capacity to manipulate the cell cycle pharmacologically. The desired ultimate outcome is to discover new classes of drugs that regulate the cell cycle for the treatment of important diseases that are not treated satisfactorily today.

As a result of the far-reaching developments made in our understanding of the regulation of the cell cycle, and the importance of the cell cycle in the pathogenesis of disease, we believed that the time was appropriate to review many of the major developments in this field. While much of the focus of research on cell cycle regulation has, and continues to be, related to the field of cancer, it is important to note that cellular proliferation resulting from activation of the cell cycle is also a key process involved in hematopoiesis, angiogenesis, inflammation and cytokine biosynthesis, organ remodeling and apoptosis. Accordingly, we have focused in this volume on the critical role that regulation of the cell cycle plays in these important pathological processes. For each of the major themes of this symposium, we enlisted opinion leaders in the field of cell cycle regulation to discuss and review the major issues related to this important topic. The proceedings of this symposium published herein represent definitive reviews of the topics that are currently believed to be among the most important in the field.

Robert R. Ruffolo, Jr.
George Poste
Brian W. Metcalf

Contributors

Alnemri, E.S.
Department of Pharmacology &
The Jefferson Cancer Institute
233 S. 10th Street
Philadelphia PA 19107
USA

Blenis, J.
Department of Cell Biology
Harvard Medical School
25 Shattuck Street
Boston MA 02115
USA

Brown, L.
Biochemistry of the Cell Nucleus
Laboratory
Imperial Cancer Research Fund
PO Box 123
44 Lincoln's Inn Fields
London WC2A 3PX
UK

Da Silva, L.
Biological Carcinogenesis &
Development Progam
Science Applications International
Corporation Frederick
National Cancer Institute
Frederick Cancer Research &
Development Center
Frederick MD 21702-1201
USA

Duhe, R.J.
Biological Carcinogenesis &
Development Program
Science Applications International
Corporation
Frederick National Cancer Institute
Frederick Cancer Research &
Development Center
Frederick MD 21702-1201
USA

Earnshaw, W.C.
Department of Cell Biology &
Anatomy
Johns Hopkins School of Medicine
725 North Wolfe Street
Baltimore MD 21205
USA

Erwin, R.
Biological Carcinogenesis &
Development Program
Science Applications International
Corporation
Frederick National Cancer Institute
Frederick Cancer Research &
Development Center
Frederick MD 21702-1201
USA

Evan, G.I.
Biochemistry of the Cell Nucleus
Laboratory
Imperial Cancer Research Fund
PO Box 123
44 Lincoln's Inn Fields
London WC2A 3PX
UK

Evans, G.A.
Biological Carcinogenesis &
Development Program
Science Applications International
Corporation Frederick
National Cancer Institute
Frederick Cancer Research &
Development Center
Frederick MD 21702-1201
USA

Farrar, W.L.
Laboratory of Molecular
Immunoregulation
Cytokine Molecular Mechanisms
Section
National Cancer Institute
Frederick Cancer Research &
Development Center
Frederick MD 21702
USA

Fernandes-Alnemri, T.
Department of Pharmacology &
The Jefferson Cancer Institute
233 S. 10th Street
Philadelphia PA 19107
USA

Foster, D.
Department of Microbiology &
Immunology
State University of New York
Health Science Center at Syracuse
750 E. Adams
Syracuse NY 13210
USA

Goldman, R.D.
Department of Cell and Molecular
Biology
Northwestern University Medical
School
303 East Chicago Avenue
Chicago IL 60611
USA

Grammer, T.C.
Department of Cell Biology
Harvard Medical School
25 Shattuck Street
Boston MA 02115
USA

Howard, O.M.Z.
Biological Carcinogenesis &
Development Program
Science Applications International
Corporation Frederick
National Cancer Institute
Frederick Cancer Research &
Development Center
Frederick MD 21702-1201
USA

Ihle, J.N.
St. Jude Children's Research Hospital
332 North Lauderdale
Memphis TN 38101
USA

Jackson, J.R.
Department of Immunopharmacology,
UW-532
SmithKline Beecham Pharmaceuticals
709 Swedeland Road
PO Box 1539
King of Prussia PA 19406
USA

Kauffmann-Zeh, A.
Biochemistry of the Cell Nucleus
Laboratory
Imperial Cancer Research Fund
PO Box 123
44 Lincoln's Inn Fields
London WC2A 3PX
UK

Kaufmann, S.H.
Division of Oncology Research
Mayo Clinic
200 First Street S.W.
Rochester MN 55905
USA

Kirken, R.A.
Laboratory of Molecular
Immunoregulation
Cytokine Molecular Mechanisms
Section
National Cancer Institute
Frederick Cancer Research &
Development Center
Frederick MD 21702
USA

Kumar, S.
Department of Cellular Biochemistry
SmithKline Beecham Pharmaceuticals
PO Box 1539
King of Prussia PA 19406
USA

Lansdorp, P.M.
Terry Fox Laboratoy
601 West 10th Avenue
Vancouver BC V5Z 1L3
Canada

Lazebnik, Y.A.
Cold Spring Harbor Laboratory
PO Box 100
Cold Spring Harbor NY 11724
USA

Lee, J.C.
Department of Cellular Biochemistry
SmithKline Beecham Phamaceutical
PO Box 1539
King of Prussia PA 19406
USA

Malabarba, M.G.
Biological Carcinogenesis &
Development Program
Science Applications International
Corporation Frederick
National Cancer Institute
Frederick Cancer Research &
Development Center
Frederick MD 21702-1201
USA

McCarthy, N.
Biochemistry of the Cell Nucleus
Laboratory
Imperial Cancer Research Fund
PO Box 123
44 Lincoln's Inn Fields
London WC2A 3PX
UK

Moir, R.D.
Department of Cell and Molecular
Biology
Northwestern University Medical
School
303 East Chicago Avenue
Chicago IL 60611
USA

Poirier, G.G.
Poly (ADP-Ribose) Metabolism
Group
Department of Molecular
Endocrinology
Centre Hospitalier de l'Université
Laval Reseach Center and
Laval University
Sainte-Foy
Quebec Canada G1V 4G2

Reed, J.C.
The La Jolla Cancer Research
Foundation
Cancer Research Center
10901 N. Torrey Pines Road
La Jolla CA 92037
USA

Taffet, S .M.
Department of Microbiology &
Immunology
State University of New York
Health Science Center at Syracuse
750 E. Adams
Syracuse NY 13210
USA

Takahashi, A.
Department of Cell Biology &
Anatomy
Johns Hopkins School of Medicine
725 North Wolfe Street
Baltimore MD 21205
USA

Ulrich, E.
Biochemistry of the Cell Nucleus
Laboratory
Imperial Cancer Research Fund
PO Box 123
44 Lincoln's Inn Fields
London WC2A 3PX
UK

Whyte, M.
Biochemistry of the Cell Nucleus
Laboratory
Imperial Cancer Research Fund
PO Box 123
44 Lincoln's Inn Fields
London WC2A 3PX
UK

Winkler, J.D.
Department of Immunopharmacology
UW-532
SmithKline Beecham Pharmaceuticals
709 Swedeland Road
PO Box 1539
King of Prussia PA 19406
USA

Young, P.R.
Department of Cellular Biochemistry
SmithKline Beecham Pharmaceuticals
PO Box 1539
King of Prussia PA 19406
USA

Cell Cycle Regulation in Hematopoietic Stem Cells

PETER M. LANSDORP

1. Introduction

The blood-forming or hematopoietic system has been described as a hierarchy of three distinct cell populations (Till *et al.*, 1964; Metcalf, 1984). The largest population consists of the most mature cell, which can be recognized by their morphology as belonging to a particular differentiation lineage. The cells in this maturation compartment have a very limited proliferative potential (less than 5 divisions), and are derived from cells of the committed progenitor cell compartment, which have more extensive but nevertheless limited proliferation potential. Committed progenitor cells in turn are produced by a population of hematopoietic stem cells, which have been defined as pluripotent cells with self-renewal properties. Recent studies have shown extensive functional differences between fetal and adult "candidate" stem cells (Lansdorp *et al.*, 1993), which include differences in self-renewal properties (Lansdorp, in press). Developmental changes in stem cell function coincide with measurable changes in the mean copy number of telomeric repeats at the end of chromosomes in such cells (Vaziri *et al.*, 1994). To reconcile such genetic changes with historical notions about the self-renewal of stem cells, it was proposed that the replicative potential of stem cells may be finite, and self-renewal may be relative rather than absolute (Lansdorp, in press). A finite replicative potential of stem cells is in agreement with observations indicative of stem cell "exhaustion" upon serial transplantation (Schofield and Dexter, 1985; Harrison and Astle, 1982) or repeated cytotoxic therapy (Moore, 1992; Homung and Longo, 1992) as well as the observation that the great majority of stem cells (defined by almost any assay) are quiescent, non-cycling cells. The quiescent nature of stem cells in adult steady-state hematopoiesis raises several questions. What is the explanation for the differences in cell cycle characteristics between fetal liver and bone marrow stem cells? What is the mechanism responsible for the recruitment of quiescent, non-cycling stem cells into an active, cycling pool following marrow injury? How is return to the more quiescent steady-state achieved? By which mechanism(s) are stem cells activated (recruited from a quiescent state) during normal steady state hematopoiesis? This paper discusses some of the possible mechanisms and factors involved in these processes. No attempt is made to cover the extensive literature on the regulation of cell cycle properties in hematopoietic cells. Instead, it is hoped that by focusing on relatively novel insights and concepts further research in this area will be stimulated. Such research is necessary not only to advance knowledge in this fundamental aspect of stem cell biology but also because the effective manipulation of stem cell cycle properties appears to be a requirement for achieving effective retroviral-mediated transfer of genes into long-term lympho-myeloid repopulating stem cells.

2. Development of Hematopoiesis

Hematopoiesis is first observed in "blood islands" in the yolk sac. Blood island precursor cells are mesenchymal cells derived from the primitive streak region of the early blastoderm (reviewed in Zon, 1995). The mesenchymal cells appear to differentiate into clusters with peripheral endothelial cells and central hematopoietic cells, some of which rapidly differentiate along the erythrocyte lineage. Most likely primitive hematopoietic stem cells co-exist immediately adjacent to differentiated erythroid cells in these blood "island", and presumably both cell types are derived from the same yolk sac precursor. At the level of these primitive precursors, important decisions regarding the fate of daughter cells are somehow derived. Such decisions are presumably repeated over and over again in stem cells: they must contribute to the immediate future (differentiate) or contribute to the more distant future (self-renew). The mechanisms responsible for such cell fate decisions are incompletely understood, and continue to intrigue experimental hematologists and developmental biologists.

With the fusion of blood islands and the establishment of blood vessels and circulation, "definitive" hematopoietic stem cells from the yolk sac appear to move with the circulation to the fetal liver (Moore and Metcalf, 1970; Zon, 1995). It has not been proven that primitive hematopoietic cells in the yolk sac are the (only) precursors of the stem cells of the fetal liver. Recent studies suggest that, for a brief period of time, primitive hematopoietic cells may also be isolated from regions in or near the aorta of the embryos (Godin *et al.*, 1993; Medvinsky *et al.*, 1993).

Interestingly, some of the cells from yolk sac, para-aortic region, and fetal liver are capable of producing spleen colonies in irradiated adult recipients (Moore and Metcalf, 1970). This observation, together with the questionable use of spleen colony assays to measure "stem cells" (Ploemacher and Brons, 1988), is probably responsible for what appears to be an overall underappreciation of the developmental changes in the biological properties of primitive hematopoietic cells. Results showing that cells from early stages of development can function, to some extent, in adult hosts and vice versa (Fleischman and Mintz, 1984), illustrate a remarkable functional adaptive capacity of primitive hematopoietic cells. However, these studies also documented irreversible change in adult stem cells in agreement with differences in long-term repopulation potential between individual "competitive repopulating units" for fetal liver and adult bone marrow (Rebel *et al.*, in press). In earlier studies, CFU-S from different tissues were also found to be qualitatively different, in that the number of spleen colony-forming cells (CFU-S) per individual spleen colony decreased during ontogeny (Moore and Metcalf, 1970).

3. Loss of Telomeric DNA in Stem Cells

Several studies have documented that the number of telomeric $(TTAGGG)_n$ repeats in human somatic cells decreases with *in vitro* and *in vivo* cell divisions (Hastie *et al.*, 1990; Lindsey *et al.*, 1991; Counter *et al.*, 1992; Allsopp *et al.*, 1992; Vaziri *et al.*, 1993). In these studies the mean telomere length showed considerable variation between individuals and between different tissues. Studies with cultured fibroblasts have indicated that telomere length is a better predictor of replicative capacity than the actual age of the fibroblast donor (Allsopp *et al.*, 1992). Based on these studies, it appears that loss of telomeric DNA at one point may result in cell cycle exit signals involving p53 (Wynford-Thomas *et al.*, 1995), followed by apoptosis or senescence dependent on the type of cell and its environment (Lin and Benchimol, 1995). This scenario has been proposed to explain the observation by Hayflick and Moorhead that normal human fibroblasts stop dividing after a variable but finite number of doublings *in vitro* (Harley, 1991).

When DNA from fetal liver cells, umbilical cord blood cells, and adult bone marrow cells was compared, a significant difference was observed in the mean length of terminal restriction fragments (containing the $(T_2AG_3)_n$ telomere repeats) between the fetal/neonatal and adult tissues (Vaziri *et al.*, 1994). The observed loss of telomeric DNA in these hematopoietic tissues did not appear to be restricted to more mature cell, as purified $CD34^+CD38^-$ primitive progenitor cells from adult bone marrow were also found to have shorter telomeres than fetal liver cells (Vaziri *et al.*, 1994). In these studies, there also appeared to be an age-related loss of telomeric DNA. The measured rate of telomere loss in lymphocytes was calculated to reflect 0.4 stem cell doublings/year (Vaziri *et al.*, 1993), whereas the corresponding figure was 0.25 stem cell doubling/year for (limited) adult bone marrow data (Vaziri *et al.*, 1994). The considerable variation in the mean telomere length between individuals requires that these studies be extended to larger samples in order to obtain more accurate estimates of *in vivo* cell turn-over. Interestingly, leukemic blast cells were found to have significantly shorter telomeres than normal cells from the same patient (Adamson *et al.*, 1992; Yamada *et al.*, 1995). All of these findings are in support of proliferation-dependent losses of telomeric DNA in cells of hematopoietic origin. Measurements of telomere length in cells could thus be used to assess the proliferative history and proliferation potential of cells.

4. The Intrinsic Time-table Model

In order to reconcile a finite replicative potential of stem cells with developmental changes in stem cell function, telomere shortening and transplantation data, a simplified model of stem cell biology, the Intrinsic Time-table (IT) model, was recently proposed (Lansdorp, 1994; Lansdorp, in press). In this model, self-renewal is relative, under strict developmental control, and predicable at a population level but unpredictable at the level of single cells. The key assumption on which IT is based is that life-long production of mature blood cells is derived from primordial stem cells with a **limited** replication potential. It is not a priori impossible that a limited number of cell divisions are at least in theory sufficient to meet the daily requirements for mature blood cells (estimated to be in the order of 10^{12} cells/day). According to a simple calculation, a replication potential of only seventy-five divisions in a stem cell could in theory give rise to 4×10^{21} blood cells, which would in fact represent a large excess relative to maximum requirements for lifelong steady-state hematopoiesis (estimated to be in the order of 10^{16}–10^{17} cells/life). Such an excess would presumably also be sufficient to explain regeneration and transplantation data. The assumption that stem cells have a finite replication potential assumes that few cells are wasted in the hematopoietic system (i.e., by apoptosis, selection, or competition for limited stem cell "niches"). The eonomical use of a finite replication potential in stem cells is compatible with the observation that the large majority of adult stem cells (defined by almost any assay) are quiescent, non-cycling cells. The final assumption in IT is that the probability of a self-renewal division as well as the turnover speed (generation or cell cycle time) of stem cells is under tight developmental control. By somehow arranging for a self-renewal probability > 0.5 during fetal development, the organism ensures that the number of stem cells increases with the increase in cell mass, while also ensuring increased production of mature blood cells by increasing the numbers of stem cells that irreversibly commit to the various differentiation pathways. The limited turnover in adult stem cells furthermore ensures that the limited replication potential of the stem cells is not exhausted during a normal life-time. IT leads to several predictions that can be tested and cautions against common notions about the self-renewal properties of hematopoietic stem cells. Several questions related to the mechanisms in control of stem cell cycle characteristics are raised by the model and need to be answered: What are the mechanisms responsible for the changes in cell cycle properties at various stages of development? To what extent are factors in the microenvironment (i.e., cytokines) capable of modulating self-renewal, differentiation, and the cell cycle status of stem cells? How is replicative exhaustion prevented in steady state and regenerative hematopoiesis? Some of the intrinsic and extrinsic factors that may be involved in these processes are discussed below.

5. Intrinsic Factors Involved in Stem Cell Cycling

Although factors in the microenvirortment, cell-cell interactions, and cytokines undoubtedly play an important role in modulating cell cycle properties of stem cells (see below), results from several experiments argue against such extrinsic factors being in sole control of stem cell cycle characteristics. Studies with purified "candidate" stem cells have shown that the majority of such cells from adult bone marrow (removed from a potentially "inhibitory" microenvironment) do not proliferate within a week in response to a mixture of IL-3, IL-6, Steel Factor, and erythropoietin (Lansdorp *et al.*, 1993; Lansdorp and Dragowska, 1993). Indeed, some of the cells in such cultures survived in a quiescent state during the 5-wk period of the experiments (Lansdorp and Dragowska, 1993). The same experimental conditions induced a rapid proliferative response in more committed progenitors. These observations argue in favor of intrinsic differences in growth factor requirements between committed progenitors and their precursor; this was also suggested in previous studies (Leary *et al.*, 1992). One factor that could be related to differences in growth factor responsiveness between colony-forming cells and their precursors is the higher expression of hematopoietic cell phosphatase (HCP, or PTPIC) in the most primitive cells (Sauvageau *et al.*, 1993). This phosphatase associates with and may interfere with signaling through c-kit and other cytokines (Yi and Ihle, 1993). High levels of HCP or similar molecules in the most primitive hematopoietic cells may decrease responsiveness to stimulatory cytokine signals below a threshold required for activation. Such signal transduction suppression may allow the maintenance of stem cells in a quiescent state for long periods of time. Following this theory, it could be argued that one of the primary hematopoietic defects in the (HCP deficient) Motheaten mouse (Tsui and Tsui, 1994) could be the uncontrolled and excessive production of committed progenitor cells from a population of stem cells that have lost a key molecule required for their maintenance in a quiescent state. The continuous cycling of stem cells in Motheaten mice could result in large numbers of progenitor cells (i.e., of the B cell lineage) which may get lost due to lack of a suitable microenvironment for further maturation. Both stem cell exhaustion and inappropriate loss of progenitor cells could contribute to the pathology in Motheaten mice.

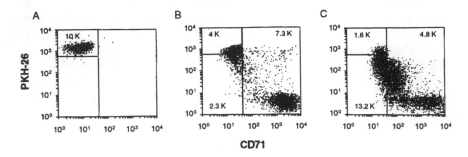

Figure 1. Stimulation of purified, quiescent human "candidate" stem cells with a CD34⁺CD45RA^lo CD71^lo phenotype (Lansdorp and Dragowska, 1993) by incubation with 150 mM of pervanadate (Evans *et al.*, 1994) for 17 hr at 37°C prior to culture of the cells in serum-free media supplemented with IL-3, IL-6, Steel factor, and erythropoietin. After purification but before pervanadate stimulation, cells were labeled with PKH26 (A), allowing the analysis of the proliferation history of the cells at later time points. At day 15 the pervanadate stimulated (C) and control cells (B) were reanalyzed. The PKH26 fluorescence is shown plotted against the CD71 (transferrin receptor) fluorescence among the (gated) cells that continued to express CD34 as well as the number of cells recovered in the various fractions.

The premise that increased levels of protein phosphorylation in stem cells may lead to their activation and proliferation is supported by results of experiments in which purified human "candidate" stem cells were treated with the phosphatase inhibitor pervanadate. As shown in Fig. 1, treatment with pervanadate resulted in an increased proliferative response of the purified (quiescent) cells to erythropoietin, Steel factor, IL-3, and IL-6.

In view of the changes in stem cell turnover during development (Lansdorp *et al.*, 1993), some of the intrinsic factors involved in stem cell cycle regulation could be developmentally regulated. One possibility is that telomere shortening itself is involved in such regulation (Fig. 2). Various models in which gradual telomere shortening could lead to suppression of cell cycle entry and progression are discussed in the legend of Fig. 2. The idea that telomere shortening could be involved in the regulation of stem cell cycling during normal hematopoiesis represents a departure from current models implicating telomere shortening in cellular senescence and aging (Wright and Shay, 1992; Harley, 1991). A key and as yet unproven assumption in this idea is that telomere shortening could be involved in the **reversible** inhibition of stem cell cycling. Low levels of the enzyme telomerase (capable of elongating telomeres) have been found in the most primitive hematopoietic cells (Broccoli *et al.*, 1995; Hiyama *et al.*, 1995; Chiu *et al.*, submitted). A gradual (slow) extension of telomeres could counter cell cycle exit signals mediated via p53 and telomere shortening. Futhermore, such suppression of stem cell cycling properties has to be reversible in order to explain stem cell cycling following stimulation (e.g., after marrow injury).

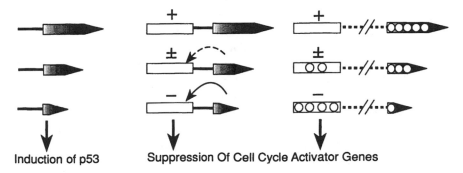

Induction of p53   **Suppression Of Cell Cycle Activator Genes**

Figure 2. Models implicating the loss of telomeric repeats in the regulation of the cell cycle in hematopoietic stem cells. Loss of telomeric repeats upon stem cell replication could directly trigger cell cycle exit via p53 (Fig. 1, left), as a critical number of repeats may be required for preventing activation of this "DNA repair" pathway in normal and malignant cells (Wynford-Thomas *et al.*, 1995). Low levels of telomerase or a shift in the balance between signals that trigger and suppress cell cycle progression could allow re-entry of stem cells into the proliferative pool at a later stage. Indirect suppression of "cell cycle activator" genes with a telomeric location (Fig. 1, middle), could result from extension of telomeric heterochromatin into such genes as the number of telomere repeats at specific chromosomes decrease (Wright and Shay, 1992). Finally, (Fig. 1, right), the activation of "cell cycle suppressor genes" (Aparicio and Gottschling, 1994) or the suppression of "cell cycle activation genes" could result from the binding to critical regulatory regions of such genes by proteins with affinity for both telomeric repeats and such regulatory regions (Kilian *et al.*, 1995).

Careful studies of telomerase expression and telomere shortening in individual stem cells in relation to the expression of factors that are now known to regulate the cell cycle could be used to examine the possible involvement of telomeres in the reversible inhibition of stem cell cycling.

6. *Extrinsic Factors Involved in Stem Cell Cycling*

It is generally thought that both positive and negative signals (i.e., cytokines in the microenvironment) are involved in the regulation of stem cell cycle properties. If these were the only factors, it is clear that interaction between such positive and negative signals would need to be delicately balanced in order to avoid hyper- or hypoplasia. It is possible that the role of such extrinsic factors relative to intrinsic factors such as those discussed above have been overestimated, and that stem cells in a suitable microenvironment will "autoregulate" their cycling properties (default to quiescence for cells from adult tissue) by primarily intrinsic mechanisms.

Among the many positive stimulators of stem cells, the ligand for flt3/flk-2 (Lyman *et al.*, 1993; Hannum *et al.*, 1994) deserves special mention. This molecule is a powerful synergistic factor that stimulates the proliferation of purified "candidate" stem cells from human and murine bone marrow and fetal liver. Indeed, when added to cultures of human stem cell "candidates" in the previously described mixture of cytokines (Fig. 1), flt3 ligand induced a stronger proliferative response in the majority of cells than was observed using pervanadate (Lansdorp & Dragowska, unpublished observations). The requirement for multiple positive stimulatory signals and the synergistic effects of such factors on the activation of primitive cells is well documented (Leary *et al.*, 1992; Lansdorp and Dragowska, 1992). Combinations of growth factors at supra-physiological concentrations appear to be particularly powerful stimulators of stem cell cycling. Several cytokines have been proposed in the negative regulation of stem cell cycling including TGF-β1 (Ohta *et al.*, 1987; Keller *et al.*, 1988), a tetrapeptide inhibitor (Lenfant *et al.*, 1989), M1P1α (Graham *et al.*, 1990), and interferony γ (Snoeck *et al.*, 1994). It is possible that intrinsic "unresponsiveness" (discussed above) combined with these or other negative factors in the microenvironment, is responsible for the maintenance of the majority of adult stem cells during steady state hematopoiesis in a quiescent state. Studies in which CDw9O (Thy-1) on the surface of primitive hematopoietic cells was cross-linked resulted in a significant and specific inhibition in the formation of primitive hematopoietic colonies (Mayani and Lansdorp, 19914), suggesting that Thy-1 may be one of the many cell surface molecules that are undoubtedly involved in the extrinsic regulation of the cell cycle in stem cells.

7. Concluding Remarks

Studies of the mechanism involved in the regulation of stem cell cycling are being pursued at various levels will eventually have to be integrated. Insights into the pathways and factors at each level are far from complete. This can be illustrated by a discussion of the possible mechanisms responsible for the *in vivo* activation of stem cells from a quiescent state following marrow injury (i.e., by irradiation or cytotoxic therapy). Are increases in positive regulators responsible for this shift, or is a down-regulation of inhibitory signals involved? Are factors such as increased "space" allowing perhaps for increased mobility, involved in switching the balance from suppression to activation of stem cells? Perhaps the activation of stem cells requires multiple steps, such as the removal from a primarily inhibitory microenvironment (achieved by "mobilization" of stem cells following marrow injury or cytokine treatment) followed by the homing of the mobilized cells into a less suppressive microenvironment that supports the proliferation of the activated cells. These questions and scenarios for stem cell activation following marrow injury must be addressed and examined. With the availability of various positive and negative regulators, as well as the stem cell themselves in highly purified form, many of the questions in this field could be answered in the near future. By combining the data from such *in vitro* studies with functional studies of stem cell cycling in various genetically-deficient mice, a complete picture of stem cell cycle control mechanisms should eventually emerge. Such a picture will no doubt point to ways in which the cell cycle properties of normal stem cells can be manipulated to achieve therapeutic benefits.

Acknowledgments

This work was supported by NIH grant A129524 and by a grant from the National Cancer Institute of Cananda. Visia Dragowska is thanked for expert technical assistance, and Colleen MacKinnon for typing the manuscript.

References

Adamson, D.J.A., King, D.J., and Haites, N.E., 1992, Significant telomere shortening in childhood leukemia, *Cancer Genet. Cytogenet.*, **61**:204–206.

Allsopp, R.C., Vaziri, H., Patterson, C., Goldstein, S., Younglai, E.V., Futcher, A.B., Greider, C.W., and Harley, C.B., 1992, Telomere length predicts repticative capacity of human fibroblasts, *Proc. Natl. Acad. Sci. USA*, **89**:10114–10118.

Aparicio, O.M., and Gottschling, D.E., 1994, Overcoming telomeric silencing: a trans-activator competes to establish gene expression in a cell cycle-dependent way, *Genes Dev.*, **8**:1133–1146.

Broccoli, D., Young, J.W., and de Lange, T., 1995, Telomerase activity in normal and malignant hematopoietic cells, *Proc. Natl. Acad. Sci. USA*, **92**:9082–9086.

Chiu, C-P., Dragowska, W., Kim, N.W., Vaziri, H., Yui, J.S., Thomas, T.E., Harley, C.B., and Lansdorp, P.M., Differential expression of telomerase activity in hematopoietic progenitors from adult human bone marrow, *Stem Cells* (Submitted).

Counter, C.M., Avilion, A.A., LeFeuvre, C.E., Stewart, N.G., Greider, C.W., Harley, C.B., and Bacchetti, S., 1992, Telomere shortening associated with chromosome instability is arrested in immortal cells which express telomerase activity, *EMBO J.*, **11**:1921–1929.

Evans, G.A., Garcia, G.G., Erwin, R., Howard, O.M.Z., and Farrar, W.L., 1994, Pervanadate simulates the effects of interieukin-2 (IL-2) in human T cells and provides evidence for the activation of two distinct tyrosine kinase pathways by IL-2, *J. Biol. Chem.*, **269**:23407–23412.

Flcischman, R.A., and Mintz, B., 1984, Development of adult bone marrow stem cells in H-2-compatible and -incompatible mouse fetuses, *J. Exp. Med.*, **159**:731–745.

Godin, I.E., Garcia-Porrero, J.A., Coutinho, A., Dieterlen-Lievre, F., and Marcos, M.A.R., 1993, Para-aortic splanchnopleura from early mouse embryos contains B1a cell progenitors, *Nature*, **364**:67–69.

Graham, G.J., Wright, E.G., Hewick, R., Wolpe, S.D., Wilkie, N.M., Donaldson, D., Lorimore, S., and Pragnell, I.B., 1990, Identification and characterization of an inhibitor of haemopoietic stem cell proliferation, *Nature*, **344**:442–444.

Hannum, C., Culpepper, J., Campbell, D., McClanahan, T., Zurawski, S., Bazan, J.F., Kastelein, R., Hudak, S., Wagner, J., Mattson, J., Luh, J., Duda, G., Martina, N., Peterson, D., Menon, S., Shanafelt, A., Muench, M., Kelner, G., Namikawa, R., Rennick, D., Roncarolo, M.G., Zlotnik, A., Rosnet, O., Dubreuil, P., Birnbaum, D., and Lee, F., 1994, Ligand for FLT3/FLK2 receptor tyrosine kinase regulates growth of haematopoietic stem cells and is encoded by variant RNAS, *Nature*, **368**:643–648.

Harley, C.B., 1991, Telomere loss: mitotic clock or genetic time bomb? *Mutat. Res.*, **256**:271–282.

Harrison, D.E., and Astle, C.M., 1982, Loss of stem cell repopulating ability upon transplantation. Effects of donor age, cell number, and transplantation procedure, *J. Exp. Med.*, **156**:1767–1779.

Hastie, N.D., Dempster, M., Dunlop, M.G., Thompson, A.M., Green, D.K., and Alishire, R.C., 1990, Telomere reduction in human colorectal carcinoma and with ageing, *Nature*, **346**:866–868.

Hiyama, K., Hirai, Y., Kyoizumi, S., Akiyama, M., Hiyama, E., Piatyszek, M.A., Shay, J.W., Ishioka, S., and Yamakido, M., 1995, Activation of telomerase in human lymphocytes and hematopoietic progenitor cells, *J. Immunol.*, **155**:3711–3715.

Hornung, R.L., and Longo, D.L., 1992, Hematopoietic stem cell depletion by restorative growth factor regimens during repeated high dose cyclophosphamide therapy, *Blood*, **80**:77–83.

Keller, J.R., Mantel, C., Sing, G.K., Ellingsworth, L.R., Ruscetti, S.K., and Ruscetti, F.W., 1988, Transforming growth factor β1 selectively regulates early murine hematopoietic progenitors and inhibits the growth of IL-3-dependent myeloid leukemia cell lines, *J. Exp. Med.*, **168**:737–750.

Kilian, A., Stiff, C., and Kleinhofs, A., 1995, Barley telomeres shorten during differentiation but grow in callus culture, *Proc. Natl. Acad. Sci. USA*, **92**:9555–9559.

Lansdorp, P.M., 1994, Properties of purified stem cells, *Exp. Hematol.*, **22**:714.

Lansdorp, P.M., Self-renewal of stem cells, *Marrow Transplant Reviews* (in press).

Lansdorp, P.M. and Dragowska, W., 1992, Long-term erythropoiesis from constant numbers of CD34+ cells in serum-free cultures initiated with highly purified progenitor cells from human bone marrow, *J. Exp. Med.*, **175**:1501–1509.

Lansdorp, P.M. and Dragowska, W., 1993, Maintenance of hematopoiesis in serum-free bone marrow cultures involves sequential recruitment of quiescent progenitors, *Exp. Hematol.*, **21**:1321–1327.

Lansdorp, P.M., Dragowska, W., and Mayani, H., 1993, Ontogeny-related changes in proliferative potential of human hematopoietic cells, *J. Exp. Med.*, **178**:787–791.

Leary, A.G., Zeng, H.Q., Clark, S.C., and Ogawa, M., 1992, Growth factor requirements for survival in G_0 and entry into the cell cycle of primitive human hemopoietic progenitors, *Proc. Natl. Acad. Sci. USA*, **89**:4013–4017.

Lenfant, M., Wdzieczak-Bakala, J., Guittet, E., Prome, J., Scotty, D., and Frindel, E., 1989, Inhibitor of hematopoietic pluripotent stem cell proliferation: purification and determination of its structure, *Proc. Natl. Acad. Sci. USA*, **86**:779–782.

Lin, Y., and Benchimol, S., 1995, Cytokines inhibit p53-mediated apoptosis but not p53-mediated G_1 arrest, *Mol. Cell Biol.*, **15**:6045–6054.

Lindsey, J., McGill, N.I., Lindsey, L.A., Green, D.K., and Cooke, H.J., 1991, *In vivo* loss of telomeric repeats with age in humans, *Mutat. Res.*, **256**:45–48.

Lyman, S.D., James, L., Vanden Bos, T., de Vries, P., Brasel, K., Gliniak, B., Hollingsworth, L.T., Picha, K.S., McKenna, H.J., Splett, R.R., Fletcher, F.A., Maraskovsky, E., Farrah, T., Foxworthe, D., Williams, D.E., and Beclanann, M.P., 1993, Molecular cloning of a ligand for the flt3/flk-2 tyrosine kinase receptor: A proliferative factor for primitive hematopoietic cells, *Cell*, **75**:1157–1167.

Mayani, H., and Lansdorp, P.M., 1994, Thy-1 expression is linked to functional properties of primitive hematopoietic progenitor cells from human umbilical cord blood, *Blood*, **83**:2410–2417.

Medvinsky, A.L., Samoylina, N. L., Muller, A.M., and Dzierzak, E.A., 1993, An early pre-liver intra-embryonic source of CFU-S in the developing mouse, *Nature*, **364**:64–67.

Metcalf, D., 1984, *The Hemopoietic Colony Stimulating Factors*. Elsevier, Amsterdam.

Moore, M.A.S., 1992, Does stem cell exhaustion result from combining hematopoietic growth factors with chemotherapy? If so, how do we prevent it? *Blood*, **80**:3–7.

Moore, M.A.S., and Metcalf, D., 1970, Ontogeny of the haemopoietic system; yolk sac origin of *in vivo* and *in vitro* colony forming cell in the developing mouse embryo, *Br. I. Haematol.*, **18**:279.

Ohta, M., Greenberger, J.S., Anklesaria, P., Bassols, A., and Massague, J., 1987, Two forms of transforming growth factors distinguished by multipotential haemopoietic progenitor cells, *Nature*, **329**:539–545.

Ploemacher, R.E., and Brons, N.H.C., 1988, Isolation of hemopoietic stem cell subsets from murine bone marrow: II. Evidence for an early precursor of day-12 CFU-S and cells associated with radioprotective ability, *Exp. Hematol.*, **16**:27–32.

Rebel, V.I., Miller, C.L., Eaves, C.J., and Lansdorp, P.M., The repopulation potential of fetal liver hematopoietic stem cells in mice exceeds that of their adult bone marrow counterparts, *Blood* (in press).

Sauvageau, G., Deflaire, G., Lansdorp, P., Dragowska, V., and Humphries, R.K., 1993, Expression of proteintyrosine phosphatases in primitive human hemopoietic cells. *Blood*, **82**, :104a.

Schofield, R., and Dexter, T.M., 1985, Studies on the self-renewal ability of CFU-S which have been serially transferred in long-term culture or *in vivo*, *Leuk. Res.*, **9**:305–313.

Snoeck, H-W., van Bockstaele, D.R., Nys, G., Lenjou, M., Lardon, F., Haenen, L., Rodrigus, I., Peetermans, M.E., and Berneman, Z.N., 1994, Interferon γ selectively inhibits very primitive CD34^{2+}CD38$^-$ and not more mature CD34$^+$CD38$^+$, human hematopoietic progenitor cells, *J. Exp. Med.*, **180**:1177–1182.

Till, J.E., McCulloch, E.A., and Siminovitch, L., 1964, A stochastic model of stem cell proliferation, based on the growth of spleen colony-forming cells, *Proc. Natl. Acad. Sci. USA*, **51**:29–36.

Tsui, F.W.L., and Tsui, H.W., 1994, Molecular basis of the *motheaten* phenotype, *Immunol. Rev.*, **138**:185–206.

Vaziri, H., Schachter, F., Uchida, I., Wei, L., Zhu, X., Effros, R., Cohen, D., and Harley, C.B., 1993, Loss of telomeric DNA during aging of normal and trisomy 21 human lymphocytes, *Am. J. Hum. Genet.*, **52**:661–667.

Vaziri, H., Dragowska, W., Allsopp, R.C., Thomas, T.E., Harley, C.B., and Lansdorp, P.M., 1994, Evidence for a mitotic clock in human hematopoietic stem cens: Loss of telomeric DNA with age, *Proc. Natl. Acad. Sci. USA*, **91**:9857–9860.

Wright, W.E., and Shay, J.W., 1992, Telomere positional effects and the regulation of cellular senescence, *TIG*, **8**:193–197.

Wynford-Thomas, D., Bond, J.A., Wyllie, F.S., and Jones, C.J., 1995, Does telomere shortening drive selection for p53 mutation in human cancer? *Mol. Carcinogenesis*, **12**:119–123.

Yamada, O., Oshimi, K., Motoji, T., and Mizoguchi, H., 1995, Telomeric DNA in normal and leukemic blood cells, *J. Clin. Invest.*, **95**:1117–1123.

Yi, T., and Ihle, J.N., 1993, Association of hematopoietic cell phosphatase with c-Kit after stimulation with c-Kit ligand, *Mol. Cell Biol.*, **13**:3350–3358.

Zon, L.I., 1995, Developmental biology of hematopoiesis, *Blood*, **86**:2876–2891.

2

Coupled Control of Cell Proliferation and Cell Viability by Oncogenes and Cytokines

GERARD I. EVAN*, NICOLA McCARTHY, MOIRA WHYTE,
LAMORNA BROWN, ANDREA KAUFFMANN-ZEH
and EUGEN ULRICH

1. The Neoplasia Problem

The evolution of clonal mutants of somatic cells that have acquired a growth advantage over their sibling cells presents a continuous and significant neoplastic risk to multicellular organisms. In principle, any somatic cell that acquires a growth advantage should spontaneously outgrow its siblings, spread, invade, and so form a tumor. This risk is greatly exacerbated in organisms such as vertebrates due to their greater physical size (more potential cell targets), longevity (more time for mutations to occur), and because many of their tissues continuously or sporadically proliferate throughout life (sustained mutagenic risk). Thus, cancer can be viewed as an inevitable consequence of natural selection within the soma. Nonetheless, cancer affects only 1 in 3 humans, and as cancer is a clonal disease that arises through expansion of a single affected cell, this implies that the cancer cell also arises only in 1 in 3 persons, out of some 10^{14}–10^{15} somatic cells and some 10^{18} cell divisions. The cancer cell is, therefore, extremely rare, and the disease of cancer is frequently only because humans comprise so many cells. The extreme rarity of the cancer cell implies the existence of powerful mechanisms that suppress neoplasia.

*Correspondence: Gerard I. Evan, Biochemistry of the Cell Nucleus Laboratory, Imperial Cancer Research Fund, PO Box 123, 44, Lincoln's Inn Fields, London WC2A 3PX, UK. Tel: +44 171 269 3439; Fax: +44 171 269 3581; E-mail: evan@europa.lif.icnet.uk

Complex metazoans thus face the conundrum of how to generate and maintain tissue architecture and integrity, which requires massive cell proliferation both during development and throughout life, while at the same time tightly suppressing the evolution of faster growing clonal variants. Recently, it has become clear that part of the solution to this problem may involve coupling of the processes of cell proliferation and programmed cell death (apoptosis). Excess cell proliferation triggers a "compensating" increase in apoptosis, leading to an innate balance in tissue mass. Thus, cancer requires not only elevated cell proliferation but also the suppression of programmed cell death. This important conclusion arises from the surprising observation that dominant oncogenes that promote cell proliferation are also potent triggers of apoptosis.

1.1. Cell Proliferation and Cell Death are Coupled Processes

Proto-oncogenes encode essential components of signal transduction pathways that regulate cell proliferation: their inappropriate activation promotes cellular transformation and is implicated in virtually all human tumors. Recently, many dominant oncogenes have been shown to possess the unexpected biological property of triggering apoptosis. This has been most graphically demonstrated in the case of the adenovirus transforming protein E1A and the proto-oncogene c-*myc*.

E1A is the principal adenovirus early gene required to drive host cell proliferation necessary for virus replication. In mutant viruses that lack a functional version of the other adenovirus early gene, E1B, E1A remains an effective mitogenic agent but becomes also a potent trigger of apoptosis (White *et al.*, 1991). Elegant studies have demonstrated that a major function of the E1B polypeptides is to suppress apoptosis induced when E1A forces the host cell to survive and permit virus replication (Debbas and White, 1993). The regions of the E1A protein required for induction of apoptosis coincide completely with those regions necessary for promoting cell proliferation, arguing that similar molecular mechanisms are involved in both cell proliferation and apoptosis. These critical domains mediate interaction with the cellular proteins p105^{rb} and p300 (Debbas and White, 1993; Rao *et al.*, 1992), both of which provide critical functions in the regulation of G1 cell cycle progression.

The c-Myc protein, like E1A, exhibits similar duality of function and acts as a potent inducer of both cell proliferation and apoptosis (Evan *et al.*, 1992). The c-Myc protein is a short-lived, sequence-specific DNA-binding protein that resembles a transcription factor of the bHLH-LZ class (Littlewood and Evan, 1994). However, very few c-Myc target genes have been identified, so it is unclear exactly how c-Myc exerts its various biological effects. Expression of c-*myc* is necessary, and in some cases appears sufficient, for cell proliferation (reviewed in Evan and Littlewood, 1993). Deregulated c-*myc* expression is virtually ubiquitous in all tumor cells (Spencer and Groudine, 1991), and is implicated in the deregulated growth control exhibited by tumor cells. As with E1A, mutational mapping of c-Myc shows a complete concordance between those domains involved in mitogenesis, apoptosis, and transcriptional activation (Amati et al., 1994; Evan et al., 1992), arguing that c-Myc regulates proliferation and apoptosis through similar molecular mechanisms.

The induction of apoptosis by c-Myc can be seen in a variety of experimental and physiological settings. *In vitro*, c-Myc acts as a potent inducer of apoptosis in serum-deprived fibroblasts (Evan *et al.*, 1992) and in interleukin-3-deprived hematopoietic cells (Askew *et al.*, 1991). The induction of apoptosis by c-Myc *in vivo* is evident in transgenic animals whose lymphocytes express deregulated c-*myc* and which exhibit elevated spontaneous apoptosis and increased sensitivity to induction of apoptosis in lymphoid organs (Dyall and Cory, 1988; Langdon *et al.*, 1988; Neiman *et al.*, 1991). In addition, substantial oncogenic synergy is observed between c-*myc* and the anti-apoptotic gene *bcl*-2 (see below) in transgenic mouse models, indicating that cell death normally acts as an important restraint to c-*myc*-induced carcinogenesis (Strasser *et al.*, 1990; Vaux *et al.*, 1988). Incidentally, *bcl*-2 also acts as an effective suppressor of E1A-induced apoptosis and can functionally replace *E1B* (Rao *et al.*, 1992). Importantly, Bcl-2 suppresses only the apoptotic actions of c-Myc and E1A, leaving their ability to promote cell proliferation unaffected (Fanidi *et al.*, 1992). Thus, the proliferative and apoptotic arms of the c-Myc and E1A pathways must be distinct at the point at which Bcl-2 acts.

The induction of apoptosis by c-Myc raises several important questions. First, is induction of apoptosis a general property of proteins that promote cell proliferation? Second, what is the molecular mechanism by which c-Myc induces apoptosis? Third, what might be the biological rationale for the toxicity of c-Myc? Fourth, if c-Myc induces both proliferation and apoptosis, what mechanism decides which of these two opposing biological programs is selected in any one cell?

Regarding the first question, there is mounting evidence that many oncoproteins that promote cell proliferation also induce apoptosis. For example, the chimaeric homeobox oncogene *E2A-PBX1*, generated during t(1;19) chromosomal translocations in some childhood leukemias, is lymphomagenic when targeted to the lymphocytes of transgenic mice. However, such mice also exhibit massive lymphocyte apoptosis during the pre-malignant phase of the disease (Dedera *et al.*, 1993). The proto-oncogene c-*fos* is also implicated in the control of apoptosis. Topographical expression of c-*fos* during mouse embryogenesis (Smeyne *et al.*, 1993) suggests that sustained expression of c-*fos* maps to regions of tissues undergoing apoptosis. Finally, the cell cycle-specific transcription factor E2F1, required for activation of G1-specific genes during G1 progression, is also a potent trigger of apoptosis under conditions (like c-Myc) of cytokine deprivation (Qin *et al.*, 1994; Wu and Levine, 1994).

The second question cannot be answered because, unfortunately, we know virtually nothing about the mechanisms by which oncoproteins promote both proliferation and apoptosis. In the case of c-Myc, it is likely that the protein acts as a transcription factor to modulate target genes involved in both processes. Nonetheless, precisely how c-Myc couples two such opposing biological processes is obscured. c-Myc might act to regulate two independent sets of genes, one of which controls proliferation while the other modulates apoptosis. Alternatively, c-Myc might regulate one set of genes whose products trigger either proliferation or apoptosis depending upon downstream contingencies. A hybrid of these two configurations is also possible. Another heterodox view might invoke non-transcriptional activities of c-Myc in either or both process. At present, we have no way to discriminate between these models.

The biological rationale for why cell proliferation and apoptosis are coupled processes, and how the decision is made as to which of these two cell fates is selected, will be discussed in the next section.

2. Dual Signal Model for c-Myc Action

Fibroblasts expressing c-Myc undergo apoptosis only when grown in medium containing low levels of serum, or if serum is replaced with platelet-poor plasma which contains few of the growth factors and cytokines present in serum. This implies that c-Myc induces apoptosis under conditions of growth factor deprivation, and we recently identified the key serum cytokines as platelet-derived growth factor (PDGF) and insulin-like growth factor-I (IGF-I) (Harrington *et al.*, 1994a). The anti-apoptotic effects of IGF-I and PDGF do not appear to be linked to either cytokine's mitogenic activities. Both cytokines block apoptosis in cells arrested with cytostatic agents, under which conditions neither factor can promote cell proliferation. Moreover, both IGF-I and PDGF also suppress apoptosis in post-commitment S/G2 phase cells, even though such cells no longer require any mitogens for cell cycle progression (Harrington *et al.*, 1994a). Thus, IGF-I and PDGF regulate cell viability by triggering specific signaling pathways in mesenchymal cells that are discrete from those that mediate mitogenesis. Further evidence indicating that IGF-I survival signaling is separate from mitogenesis is that IGF-I suppresses cell death even in cells blocked in macromolecular synthesis. Fibroblasts treated with the protein synthesis inhibitor cycloheximide and/or the RNA synthesis inhibitor actinomycin D exhibit delayed apoptosis in the presence of IGF-I, even if the IGF-I is added after the drugs (Harrington *et al.*, 1994a). Thus, suppression of apoptosis by IGF-I does not require *de novo* protein synthesis or gene expression, quite unlike mitogenesis and cell cycle progression. Rather, suppression of apoptosis by IGF-I involves the activation of pre-existing cell machinery.

DUAL SIGNAL MODEL FOR MYC-INDUCED APOPTOSIS

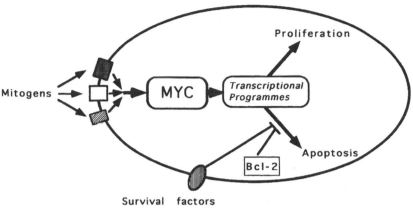

Figure 1

Based upon these observations, we have proposed a "Dual Signal" model (Fig. 1) for control of cell growth in which the processes of cell proliferation and apoptosis are obligatorily coupled, such that every cell that engages its proliferative machinery also necessarily engages a suicide "abort" program. This suicide program is only forestalled if the cell is in receipt of appropriate anti-apoptotic signals such as those provided by specific cytokines acting as "survival factors." Thus, the restricted availability of anti-apoptotic cytokines or survival factors within normal soma restrains clonal expansion by limiting cell survival. Incorporated within this "Dual Signal" hypothesis is a potent mechanism for the suppression of carcinogenesis (Evan and Littlewood, 1993; Evan *et al.*, 1992; Harrington *et al.*, 1994b). The coupling of the mitogenic and apoptotic pathways means that any oncogenic lesion that activates mitogenesis will be lethal as the affected cell and its progeny outgrow the paracrine environment enabling their survival. In effect, surveillance against neoplastic transformation is "hardwired" into the cell's proliferative machinery.

3. Cooperation between c-Myc and other Genes in Carcinogenesis

Expression of c-Myc is deregulated or elevated in virtually all human cancers. However, activated c-Myc also triggers apoptosis. This implies that tumors must, by necessity, have acquired complementary mutations that suppress apoptosis caused by growth-promoting oncogenes. What is the evidence that such mutations exist?

Deregulated expression of the oncogene *bcl*-2 is the prototypic example of an oncogenic lesion that suppresses apoptosis. *bcl*-2 was first identified as the site of reciprocal translocation on human chromosome 18 in follicular B cell lymphoma. The gene encodes a membrane associated protein, Bcl-2, present in endoplasmic reticulum, nuclear, and outer mitochondrial membranes (Krajewski *et al.*, 1993; Nakai *et al.*, 1993). Bcl-2 is widely expressed during embryonic development, but in the adult is confined to immature and stem cell populations and long-lived cells such as resting B lymphocytes and peripheral sensory neurons (Hockenbery *et al.*, 1991). Targeted expression of Bcl-2 to lymphoid cells in transgenic mice leads to an increase in the number of mature resting B cells and potentiates their longevity. Affected T cells are markedly resistant to the cytocidal effects of radiation, glucocorticoids, and anti-CD3, but thymic censorship appears normal. Bcl-2 transgenic mice go on to develop a low incidence of malignant lymphoma. However, co-expression of c-*myc* with *bcl*-2 gives rise to a markedly enhanced incidence of tumors, an incidence significantly greater than that seen with either *bcl*-2 or c-*myc* alone (reviewed in Adams and Cory, 1991). *In vitro* studies indicate that this oncogenic synergy occurs because the *bcl*-2 protein, Bcl-2, suppresses the induction of apoptosis by c-Myc (Bissonnette *et al.*, 1992; Fanidi *et al.*, 1992; Wagner *et al.*, 1993). Importantly, Bcl-2 suppresses only the apoptotic actions of c-Myc and has no effect on the ability of c-Myc to promote cell proliferation (Fanidi *et al.*, 1992). Thus, Bcl-2 does not act merely as a general antagonist of c-Myc function. Rather, it acts at a point downstream of c-Myc at which the proliferative and apoptotic arms of the c-Myc pathways are already distinct. Thus, the synergistic interaction between Bcl-2 and c-Myc is a paradigm for a particular type of oncogene cooperation in which synergy arises because one oncogene (*bcl*-2) specifically suppresses the lethality of the second (c-*myc*), without affecting the latter's capacity to promote cell proliferation. The result is a cell with markedly reduced dependence upon exogenous growth factors for either proliferation or survival. The cooperation between c-*myc* and *bcl*-2 is distinct from the classical type of oncogene cooperation; for example, that observed between c-*myc* and activated *ras* (Land *et al.*, 1983). Unlike fibroblasts expression c-*myc* and activated *ras*, fibroblasts co-expressing c-*myc* and *bcl*-2 exhibit normal flat morphology, grow only as monolayers, and fail to form typical "transformed" foci. Moreover, whereas *bcl*-2 suppresses c-*myc* induced apoptosis in low serum, activated *ras* actually exacerbates it. Thus, *ras* and *bcl*-2 co-operate with c-*myc* in very different ways (Fig. 2).

The ability of Bcl-2 to promote survival is not restricted to apoptosis induced by cytokine deprivation. Bcl-2 also makes cells less susceptible to killing by a range of cytotoxic agents, including agents such as anti-cancer drugs and radiation that act by damaging DNA integrity (Fanidi *et al.*, 1992; Fisher *et al.*, 1993; Kamesaki *et al.*, 1993; Lotem and Sachs, 1993; Miyashita and Reed, 1992; Walton *et al.*, 1993). Most significantly, Bcl-2 confers drug resistance upon cells, not by preventing the drug from inducing DNA damage, but by suppressing the suicidal

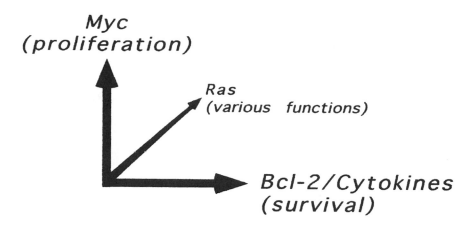

Figure 2

response of the affected cell *despite* any DNA damage that has been incurred. This constitutes a novel kind of drug resistance with particular consequences for the treatment, progression, and drug resistance of cancer. Indeed, one of the most profound implications of cell suicide has been the realization that virtually all cytotoxic agents kill cells by inducing apoptosis: "direct killing" is only observed under extremely Draconian conditions, such as immensely high levels of cytotoxic drugs or irradiation (Kerr *et al.*, 1994). This conclusion is of fundamental importance to our understanding of the mechanisms by which cytotoxic agents kill cells. Apoptosis is an innate genetic program under the aegis of a wealth of genetic and physiological parameters; in contrast, direct killing is the ineluctable consequence of sufficient cytological mayhem. If cytotoxic agents act by triggering cell suicide, suppression of apoptosis would comprise an entirely new kind of drug resistance in which cells survive not because they have evolved mechanisms to avoid drug-induced damage, but because they fail to die *even though* their DNA is damaged. In tumors containing such cells, genotoxic agents would presumably act as effective mutagens, driving the genetic diversity of their "undead" cellular targets and compounding the problems of future therapy.

bcl-2 is one member of a family of evolutionary conserved genes that modulate cell viability. These include *ced*-9 of C. *elegans* (Hengartner *et al.*, 1992); the mammalian genes *bcl*-X (Boise *et al.*, 1993), mcl-1 (Kozopas *et al.*, 1993), A1 (Lin *et al.*, 1993), bax (Oltvai *et al.*, 1993), and *bak* (Chittenden *et al.*, 1995; Farrow *et al.*, 1995; Kiefer *et al.*, 1995); and the viral proteins *BHRF1* from Epstein-Barr Virus (Pearson *et al.*, 1987) and *p19*[E1B] from adenovirus (Rao *et al.*, 1992; White, *et al.*, 1991). Functionally, these genes fall into two groups. *bcl*-2, *bcl*-X$_L$, *BHRF1*, *p19*[E1B], and *ced*-9 all suppress programmed cell death, while *bax* and *bak* appear to promote apoptosis. The antagonistic actions of differing members of the Bcl-2 protein family appear to be a result of their mutual interactions. Bcl-2 dimerizes with Bax (Oltvai *et al.*, 1993), while the favored partner for Bcl-X_L appears to be Bak (Chittenden *et al.*, 1995; Farrow 95; Kiefer *et al.*, 1995). However, it is not clear how such interactions serve to modulate cell survival. Specifically, it is not known whether Bcl-2 and Bcl-X_L activate a "survival" pathway which is quenched by sequestration with Bax and Bak, or alternatively, if Bax and Bak activate a "kill" program that is suppressed by Bcl-2 and Bcl-X_L. However, experiments indicate that ectopic expression of Bak is sufficient to trigger apoptosis in various cell types (Chittenden *et al.*, 1995), perhaps favoring the latter notion. Only by identifying the downstream targets of Bcl-2 proteins will be the molecular mechanisms underlying their actions be elucidated. One intriguing possibility is that just as Bcl-2 can act as a dominant oncogene by suppressing apoptosis and promoting survival of cells with activated *myc*, so Bak and Bax might act as tumor suppressor genes that act by triggering suicide in tumor cells.

The tumor suppressor protein p53 also plays a critical role in the regulation of cell viability through its control of apoptosis. The loss of inactivation of the p53 tumor suppressor gene is the most common single lesion in human neoplasia (Hollstein *et al.*, 1991; Nigro *et al.*, 1989; Vogelstein, 1990). The introduction of wild type p53 into p53-negative tumor cells triggers a G1 arrest (Baker *et al.*, 1990; Diller *et al.*, 1990; Martinez *et al.*, 1991), which is consistent with its biological action as a growth suppressor. Physiologically, p53 is thought to play a key role in orchestrating the cellular response to damage to genome integrity. Levels of p53 rapidly increase following DNA damage (Kastan *et al.*, 1991; Kuerbitz *et al.*, 1992; Lu and Lane, 1993), and this triggers G1 arrest, in part through trans-activation of the *Waf1/Cip1* gene which encodes an inhibitor of the G1-specific cyclin-dependent kinases (El-Deiry *et al.*, 1993; Harper *et al.*, 1993). Tumor cells lacking p53, and cells derived from mice in which p53 has been experimentally deleted, both fail to arrest in G1 following DNA damage. Instead, they enter S-phase with damaged DNA and so sustain greatly elevated risk of further mutation and carcinogenic progression (Kuerbitz *et al.*, 1992).

It is now clear that the growth-inhibiting effect of p53 is only one component of its tumor suppressive action: p53 is also a potent trigger of apoptosis following DNA damage. The re-introduction of wild-type p53 into p53-negative tumor cells triggers apoptosis (Shaw *et al.*, 1992; Yonish-Rouach *et al.*, 1991), probably due to pre-existing DNA damage present in the tumor cells. Thymocytes derived from p53[null] mice show marked radiation and genotoxic resistance (Clarke *et al.*, 1993; Lowe *et al.*, 1993). However, not all pathways to apoptosis are p53-dependent, as p53[null] thymocytes retain normal sensitivity to glucocorticoids, a cytotoxic trigger that does not involve DNA damage. Thus, p53 appears to be an essential part of a genotoxic damage pathway leading to the apoptosis of genetically damaged cells (Lane, 1992) that would otherwise present a serious neoplastic risk to the organism.

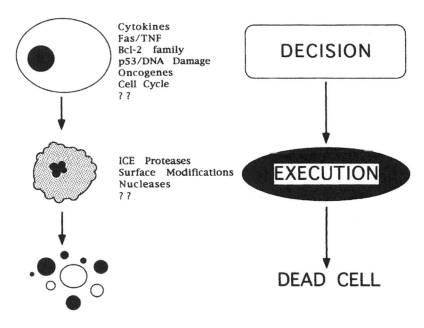

Figure 3

4. Conclusions

The fact that growth-promoting oncogenic lesions promote apoptosis, together with evidence of frequent anti-apoptotic lesions within tumor cells, attests to the importance of lesions within the cell suicide program in carcinogenesis. Nonetheless, despite obvious strong selection against apoptosis, it is intriguing that the basal apoptotic machinery itself never seems to be lost: even the most aggressive, drug-resistant, and habituated tumor cells undergo apoptosis if a potent enough trigger is administered. For example, cells that have lost p53 or that are expressing high levels of Bcl-2 or its homolgoues retain the ability to undergo apoptosis, they merely have a raised threshold at which it is triggered. Moreover, when such recalcitrant cells do eventually activate their apoptotic program, the actual process proceeds at the same rate as in sensitive cells (unpublished observations of authors). Thus, anti-apoptotic lesions within tumor cells serve to suppress activation of the program but do not appear to comprise part of the program itself (Fig. 3). Clearly, activation of the suppressed apoptotic program within tumor cells provides the basis for a totally new therapeutic strategy for treating cancer, quite unlike most existing treatment modalities that are predicated principally on the notion of the inhibiting cell proliferation. Unfortunately, at present we have little idea as to the diversity and nature of apoptosis-suppressing mutations in malignant cells. However, the identification and characterization of such lesions, and the molecular basis for their action, is certain to provide an exciting and effective set of novel targets for future therapeutic intervention in cancer.

References

Adams, J.M., and Cory, S., 1991, Transgenic models for haemopoietic malignancies, *Biochim Biophys Acta*, **1072**:9–31.

Amati, B., Littlewood, T., Evan, G. and Land, H., 1994, The c-Myc protein induces cell cycle progression and apoptosis through dimerisation with Max, *EMBO J.*, **12**:5083–5087.

Askew, D., Ashmun, R., Simmons, B., and Cleveland, J., 1991, Constitutive c-*myc*, expression in IL-3-dependent myeloid cell line suppresses cycle arrest and accelerates apoptosis, *Oncogene*, **6**: 1915–1922.

Bissonnette, R., Echeverri, F., Mahboubi, A., and Green, D., 1992, Apoptotic cell death induced by c-*myc*, is inhibited by *bcl*-2, *Nature*, **359**:552–554.

Boise, L., González-Garcia, M., Postema, C., Ding, L., Lindsten, T., Turka, L., Maxo, X., Nuñez, G., and Thompson, C., 1993, *bcl*-x, a *bcl*-2-related gene that functions as a dominant regulator of apoptotic cell death, *Cell*, **74**:597–608.

Braselmann, S., Graninger, P., and Busslinger, M., 1993, A selective transcriptional induction system for mammalian cells based on Gal4-estrogen receptor fusion proteins, *Proc. Natl. Acad. Sci. USA*, **90**: 1657–61.

Chittenden, T., Harrington, E., O'Connor, R., Evan, G. and Guild, B., 1995, Induction of apoptosis by the Bcl-2 homologue Bak, *Nature*, **374**: 733–736.

Debbas, M., and White, E., 1993, Wild-type p53 mediates apoptosis by E1A, which is inhibited by E1B, *Genes & Dev.*, **7**:546–554.

Dedera, D., Waller, E., LeBrun, D., Sen-Majumdar, A., Stevens, M., Barsh, G., and Cleary, M., 1993, Chimeric homeobox gene E2A-PBX1 induces proliferation, apoptosis and malignant lymphomas in transgenic mice, *Cell*, **74**:833–843.

Dyall, S. D., and Cory, S., 1988, Transformation of bone marrow cells from E mu-myc transgenic mice by Abelson murine leukemia virus and Harvey murine sarcoma virus, *Oncogene Res*, **2**: 403–9.

Evan, G., and Littlewood, T., 1993, The role of c-*myc*, in cell growth, *Curr. Opin. Genet. & Dev.*, **3**:44–49.

Evan, G., Wyllie, A., Gilbert, C., Littlewood, T., Land, H., Brooks, M., Waters, C., Penn, L., and Hancock, D., 1992, Induction of apoptosis in fibroblasts by c-*myc*, protein, *Cell*, **63**: 119–125.

Fanidi, A., Harrington, E., and Evans, G., 1992, Cooperative interaction between c-*myc*, and *bcl*-2 proto-oncogenes, *Nature*, **359**:554–556.

Farrow, S., White, J., Martinou, I., Raven, T., Pun, K.-T., Grinham, C., Martinou, J.-C., and Brown, R., 1995, Cloning of a novel *bcl*-2 homologue by interaction with adenovirus E1B 19K, *Nature*, **374**: 731–733.

Fernandes-Alnemri, T., Litwack, G., and Alnemri, E.S., 1994, CPP32, a novel human apoptotic protein with homology to Caenorhabditis *elegans*, cell death protein Ced-3 and mammalian interleukin-1 beta-converting enzyme, *J. Biol. Chem.*, **269**: 30761–4.

Fisher, T.C., Milner, A.E., Gregory, C.D., Jackman, A.L., Aherne, G.W., Hartley, J.A., Dive, C., and Hickman, J.A., 1993, *bcl*-2 modulation of apoptosis induced by anticancer drugs: resistance to thymidylate stress is independent of classical resistance pathways, *Cancer Res.*, **53**:3321–6.

Harrington, E., Fanidi, A., Bennett, M., and Evan, G., 1994a, Modulation of Myc-induced apoptosis by specific cytokines, *EMBO J.*, **13**: 3286–3295.

Harrington, E., Fanidi, A., and Evan, G., 1994b, Oncogenes and cell death, *Curr. Opin. Genet. Dev.*, **4**: 120–129.

Hengartner, M.O., Ellis, R.E., and Horvitz, H.R., 1992, Caenorhabditis *elegans* gene *ced*-9 protects cells from programmed cell death, *Nature*, **356**:494–9.

Hockenbery, D.M., Zutter, M., Hickey, W., Nahm, M., and Korsmeyer, S.J., 1991, Bcl-2 protein is topographically restricted in tissues and characterized by apoptotic cell death, *Proc. Natl. Acad. Sci. USA*, **88**:6961–5.

Howes, K.A., Ransom, L.N., Papermaster, D.S., Lasudry, J.G.H., Albert, D.M., and Windle, J.J., 1994, Apoptosis or retinoblastoma: alternative fates of photoreceptors expressing the HPV-16 E7 gene in the presence or absence of p53, *Genes Dev.*, **8**:1300–10.

Kamesaki, S., Kamesaki, H., Jorgensen, T., Tanizawa, A., Pommier, Y., and Cossman, J., 1993, Bcl-2 protein inhibits etoposide-induced apoptosis throught its effects on events subsequent to topoisomerase II-induced DNA strand breaks and their repair, *Cancer Research*, **53**:4251–4256.

Kastan, M.B., Onyekwere, O., Sidransky, D., Vogelstein, B., and Craig, R.W., 1991, Participation of p53 protein in the cellular response to DNA damage, *Cancer Res*, **51**: 6304–11.

Kerr, J.F., Winterford, C.M., and Harmon, B.V., 1994, Apoptosis. Its significance in cancer and cancer therapy, *Cancer*, **73**:2013–26.

Keifer, M., Brauer, M., VC, P., Wu, J., Umansky, S., Tomei, L., and Barr, P., 1995, Modulation of apoptosis by the widely distributed Bcl-2 homologue Bak, *Nature*, **374**:736–739.

Kozopas, K. M., Yang, T., Buchan, H. L., Zhou, P., and Craig, R. W., 1993, MCl1, a gene expressed in programmed myeloid cell differentiation, has sequence similarity to BCL2, *Proc. Natl. Acad. Sci. USA*, **90**:3516–20.

Krajewski, S., Tanaka, S. Takayama, S., Schibler, M.W.F., and JC., R., 1993, Investigation of the subcellular-distribution of the Bcl-2 oncoprotein – residence in the nuclear-envelope, endoplasmic-reticulum, and outer mitochondrial-membranes, *Cancer Res.*, **53**:4701–14.

Krammer, P., Behrmann, I., Daniel, P., Dhein, J., and Debatin, K.-M., 1994, Regulation of apoptosis in the immune system, *Curr. Opinion Immunol.*, **6**:279–289.

Land, H., Parada, L.F., and Weinberg, R.A., 1983, Tumorigenic conversion of primary embryo fibroblasts requires at least two cooperating oncogenes, *Nature*, **304**:596–602.

Lane, D., 1993, A death in the life of p53 [commentary], *Nature*, **362**:786.

Langdon, W.Y., Harris, A.W., and Cory, S., 1988, Growth of E mu-myc transgenic B-lymphoid cells *in vitro*, and their evolution toward autonomy, *Oncogene Res.*, **3**:271–9.

Lin, E., Orlofsky, A., Berger, M., and Prystowsky, M., 1993, Characterization of A1, a novel hemopoietic-specific early-response gene with sequence similarity to bcl-2, *J. Immunol.*, **151**:1979–1988.

Los, M., Vandecraen, M., Penning, L. C., Schenk, H., Westendorp, M., Baeuerle, P.A., Droge, W., Krammer, P.H., Fiers, W., and Schulzeosthoff, K., 1995, Requirement of an Ice/ced-3 protease for fas/apo-1-mediated apoptosis, *Nature*, **375**:81–83.

Lotem, J., and Sachs, L., 1993, Regulation by bcl-2, c-myc, and p53 of susceptibility to induction of apoptosis by heat shock and cancer chemotherapy compounds in differentiation-competent and -defective myeloid leukemic cells, *Cell Growth Differ*, **4**:41–7.

Miura, M., Zhu, H., Rotello, R., Hartwieg, E.A., and Yuan, J., 1993, Induction of apoptosis in fibroblasts by IL-1 beta-converting enzyme, a mammalian homolog of the C. elegans cell death gene ced-3, *Cell*, **75**:653–60.

Miyashita, T., and Reed, J.C., 1992, bcl-2 gene transfer increases relative resistance of S49.1 and WEH17.2 lymphoid cells to cell death and DNA fragmentation induced by glucocorticoids and multiple chemotherapeutic drugs, *Cancer Res.*, **52**:5407–11.

Myers, M.J., and White, M.F., 1993, The new elements of insulin signaling. Insulin receptor substrate-1 and proteins with SH2 domains, *Diabetes*, **42**:643–50.

Nakai, M., Takeda, A., Cleary, M.L., and Endo, T., 1993, The *bcl*-2 protein is inserted into the outer-membrane but not into the inner membrane of rat-liver mitochondria *in vitro*, *Biochem. Biophys. Res. Comms.*, **196**:233–9.

Neiman, P.E., Thomas, S.J., and Loring, G., 1991, Induction of apoptosis during normal and neoplastic B-cell development in the bursa of Fabricius, *Proc. Natl. Acad. Sci. USA*, **88**:5857–61.

Oltvai, Z., Milliman, C., and Korsmeyer, S., 1993, Bcl-2 heterodimerizes *in vivo*, with a conserved homolog, Bax, that accelerates programmed cell death, *Cell*, **74**:609–619.

Pearson, G.R., Luka, J., Petti, L., Sample, J., Birkenbach, M., Braun, D., and Kieff, E., 1987, Identification of an Epstein-Barr virus early gene encoding a second component of the restricted early antigen complex, *Virology*, **160**:151–61.

Picksley, S.M., and Lane, D.P., 1994, p53 and Rb:their cellular roles, *Curr. Opin. Cell. Biol.*, **6**:853–8.

Qin, X., Livingston, D., Kaelin, W., and Adams, P., 1994, Deregulated transcription factor e2f-1 expression leads to S-phased entry and p53-mediated apoptosis, *Proc. Natl. Acad. Sci. USA*, **91**:10918–10922.

Rao, L., Debbas, M., Sabbatini, P., Hockenberry, D., Korsmeyer, S., and White, E., 1992, The adenovirus E1A proteins induce apoptosis, which is inhibited by the E1B 19-kDa and Bcl-2 proteins, *Proc. Natl. Acad. Sci. USA*, **89**:7742–7746.

Robaye, B., Mosselmans, R., Fiers, W., Dumong, J. E., and Galand, P., 1991, Tumor necrosis factor induces apoptosis (programmed cell death) in normal endothelial cells *in vitro*, *Am. J. Pathol.*, **138**:447–53.

Shi, L., Kraut, R.P., Aebersold, R., and Greenberg, A.H., 1992, A natural killer cell granule protein that induces DNA fragmentation and apoptosis, *J. Exp. Med.*, **175**:553–66.

Smeyne, R., Vendrell, M., Hayward, M., Baker, S., Miao, G., Schilling, K., Robertson, L., Curran, T., and Morgan, J., 1993, Continuous c-*fos*, expression precedes programmed cell-death *in vivo*, *Nature*, **363**:166–169.

Spencer, C.A., and Groudine, M., 1991, Control of c-*myc*, regulation in normal and neoplastic cells, *Adv. Cancer Res.*, **56**:1–48.

Strasser, A., Harris, A. W., Bath, M. L., and Cory, S., 1990, Novel primitive lymphoid tumours induced in transgenic mice by cooperation between *myc*, and *bcl*-2, *Nature*, **348**:331–3.

Vaux, D.L., Cory, S., and Adams, J.M., 1988, *Bcl*-2 gene promotes haemopoietic cell survival and cooperates with c-*myc*, to immortalize pre-B cells, *Nature*, **335**:440–2.

Wagner, A.J., Small, M.B., and Hay, N., 1993, Myc-mediated apoptosis is blocked by ectopic expression of *bcl*-2, *Mol Cell Biol*, **13**:2432–2440.

Walton, M.I., Whysong, D., O'Connor, P.M., Hockenbery, D. Korsmeyer, S.J., and Kohn, K.W., 1993, Constitutive expression of human Bcl-2 modulates nitrogen mustard and comptothecin induced apoptosis, *Cancer Res.*, **53**:1853–61.

Wang, L., Miura, M., Bergeron, L., Zhu, H., and Yuan, J., 1994, Ich-1, an Ice/ced-3-related gene, encodes both positive and negative regulators of programmed cell-death, *Cell*, **78**:739–750.

White, E., Cipriani, R., Sabbatini, P., and Denton, A., 1991, Adenovirus E1B 19-kilodalton protein overcomes the cytotoxicity of E1A proteins, *J. Virol.*, **65**:2968–78.

White, E., Sabbatini, P., Debbas, M., Wold, W., Kusher, D., and Gooding, L., 1992, The 19-kilodalton Adenovirus E1B transforming protein inhibits programmed cell death and prevents cytolysis by tumour necrosis factor α, *Mol. Cell. Biol.*, **12**:2570–2580.

Wu, X., and Levine, A.J., 1994, p53 and E2F-1 cooperate to mediate apoptosis, *Proc. Natl. Acad. Sci. USA*, **91**:3602–6.

3

Angiogenesis and the Endothelial Cell Cycle

JEFFREY R. JACKSON and *JAMES D. WINKLER*

1. Introduction

Angiogenesis is the growth of new blood vessels from existing ones. It is an important aspect of new tissue development and tissue growth and repair. It is also a component of many diseases, including cancer, blindness, and chronic inflammation such as rheumatoid arthritis (RA) and psoriasis (Colville-Nash and Scott, 1992; Folkman and Brem, 1992; Folkman and Shing, 1992; Colville-Nash and Seed, 1993; Folkman, 1995). It has been hypothesized that angiogenesis may be such a vital aspect of these diseases that inhibiting angiogenesis may have therapeutic effects (Folkman, 1995). In order to better understand how to inhibit angiogenesis, there have been many studies of the factors that regulate this process.

One of the most interesting features of angiogenesis is that it occurs in humans in such a controled fashion. In the normal adult the endothelium is quiescent; angiogenesis is primarily observed during wound healing and in specialized tissues involved in ovulation and menstruation (D'Amore, 1992; Tobelem, 1990). Thus, endothelial cells have the ability to turn off their cell proliferation cycle and become dormant for many years, maintaining a healthy state with no apoptosis apparent. This chapter will discuss the proliferation of endothelial cells, the central cells of angiogenic blood vessels, focusing on factors that may turn the cell

Correspondence: Dr. James D. Winkler, Department of Immunopharmacology, UW-532, Smith-Kline Beecham Pharmaceuticals, 709 Swedeland Road, P.O. Box 1539, King of Prussia, PA 19406, USA. Tel: 610-270-4946; Fax: 610-270-5381; E-Mail: James_D_Winkler@SBphrd.com proliferation cycle on or off, evidence that it can be restarted, and the potential for pharmacological intervention within the cell cycle to affect angiogenesis.

2. Events of Angiogenesis

Angiogenesis is a complex multi-step process regulated by cytokines, chemokines, and growth factors (Folkman and Brem, 1992; Colville-Nash and Seed, 1993). It typically begins with the activation of vascular endothelial cells in a post-capillary venule marked by chronic vasodilation of the parent vessel. The next step requires the endothelial cells to degrade their existing basement membrane by increasing production of matrix metalloproteinases. This is followed by migration of endothelial cells though the basement membrane toward the angiogenic stimuli, forming capillary sprouts. The growth of these sprouts continues through migration of the cells in the tip and proliferation of the cells in the middle of the sprout. Lumen formation begins below the area of proliferation and migration, and probably involves the synthesis of a new basement membrane to provide mechanical support. Capillary loop formation is completed when a sprout encounters and links with another sprout, capillary, arteriole, or venule. Until it matures, the neovasculature is often leaky and poorly organized, with vessels following tortuous routes and often looping back upon themselves or shunting directly from arteriole to venule without going through the capillary bed. The consequences of this can be inefficient perfusion and continued exudation and cellular influx, leading to maintenance of the pro-angiogenic environment.

3. Regulation of Angiogenesis

The signals stimulating angiogenesis, while directed at the endothelial cells, come from other cells in the nearby tissues. Many cells are capable of producing angiogenic factors when their environment becomes hypoxic or inflammatory, including tumor cells, keratinocytes, corneal fibroblasts, synovial fibroblasts, monocytes, and macrophages. Factors capable of inducing angiogenesis can be separated into two categories: indirect, which can only induce angiogenesis *in vivo*, and direct, which can also induce proliferation, migration, and/or differentiation of endothelial cells *in vitro*.

Examples of direct factors include granulocyte- and granulocyte-macrophage colony-stimulating factors (G- and GM-CSF) (Bussolino *et al.*, 1989; Bussolino *et al.*, 1991), basic fibroblast growth factor (bFGF) (Baird and Walicke, 1989; Ernestein *et al.*, 1992), and vascular endothelial growth factor (VEGF) (Thomas, 1996; Neufeld *et al.*, 1994). VEGF has been the focus of many studies and appears to be one of the most endothelial cell-specific and unequivocal angiogenic factors. *In vitro*, VEGF causes endothelial cell proliferation and migration; *in vivo*, it is potently angiogenic and causes vascular permeability (Leung *et al.*, 1989). It has also been implicated in the angiogenesis accompanying rheumatoid arthritis (Koch *et al.*, 1994; Nagashima *et al.*, 1995). Other growth factors with some direct effects on endothelial cells include transforming growth factors α and β (TGF-α, TGF-β) and platelet derived growth factor (PDGF) (Schreiber *et al.*, 1986; Bell and Madri, 1989).

Many inflammatory mediators have both direct and indirect angiogenic activities. Prostaglandins F_1 and E_2 (PGE$_1$, PGE$_2$), tumor necrosis factor α (TNF-α), and interleukins 1, 6, and 8 (IL-1, IL-6, IL-8) have all been shown to induce angiogenesis *in vivo*. The E prostaglandins induce angiogenesis in both the cornea and chicken chorioallantoic membrane assays, with PGE$_1$ appearing more potent than PGE$_2$ in the cornea (BenEzra, 1978; Form and Auerbach, 1983; Stjernschantz *et al.*, 1989). PGE$_2$ has been shown to induce the expression of VEGF in fibroblasts (Ben-Av *et al.*, 1995). IL-6 expression coincides with early angiogenic events and thus has been implicated in angiogenesis (Motro *et al.*, 1990). It has demonstrated both stimulatory and inhibitory effects on endothelial cell growth, making its role difficult to appreciate, however; it has recently been shown to induce VEGF expression (Cohen *et al.*, 1996). IL-8, the well known inflammatory chemokine (Baggiolini and Clark-Lewis, 1992), is both chemotactic and mitogenic for endothelial cells in addition to causing angiogenesis *in vivo* in the cornea assay. IL-1α has been reported to be significantly more potent than VEGF in inducing corneal angiogenesis, requiring as little as 1 ng (BenEzra *et al.*, 1990). Indirect angiogenic activity has been demonstrated recently by the ability of IL-1 to induce VEGF expression *in vitro* (Ben-Av *et al.*, 1995; Li *et al.*, 1995). TNF-α also appears to have dose dependent effects on angiogenesis. *In vivo*, it is a potent inducer at low doses (3.5 ng in corneal implants), and an inhibitor at higher doses (Leibovich *et al.*, 1987; Fajardo *et al.*, 1992). *In vitro*, TNF induces enthothelial cell migration and the formation of capillary tubelike structures (Leibovich *et al.*, 1987).

4. Control of Endothelial Cell Proliferation

One of the key events in angiogenesis is endothelial cell proliferation. The primary characteristic of the direct angiogenic factors discussed above appears to be the ability to stimulate endothelial cell proliferation, suggesting that this is also a major point of regulation. There have been many investigations into endothelial cell proliferation, with most studies focusing on cell culture systems. It is clear from such studies that removing growth factors can block cell proliferation, and that re-adding them can restart proliferation. Among the factors shown to affect proliferation upon their removal are VEGF and bFGF (Risau, 1994). VEGF exists as various splice variants, with different properties. The smaller forms are more soluble and can diffuse over some distance to affect endothelial cells. The larger forms have higher binding affinity to components of the extracellular matrix, such as heparin, and may serve as a storage form of the growth factor, to be released after injury or during inflammatory responses. Thus, it is possible to turn off endothelial proliferation by removing required growth factors. This appears to underlie the mechanism by which a number of endogenous angiogenesis inhibitors work. For example, suramin, protamine and platelet fractor-4 have been reported to bind and inactivate growth factors such as bFGF and VEGF (Auerbach and Auerbach, 1994).

There is also a large body of work demonstrating that endothelial cells proliferate when there is a permissive environment provided by the extracellular matrix (Ingber, 1992). For example, tissue cultures plates must be coated with gelatin or another extracellular matrix material to allow cell survival and growth. Endothelial cells can proliferate and remodel in different ways when exposed to various mixtures of extracellular matrix materials (Haralabopoulos et al., 1994). Recent reports have shown that proliferating and migrating endothelial cells may require the presence of specific adhesion molecules in the extracellular matrix. It has been shown, using antibodies directed against the integrin $\alpha v\beta 3$, that when this adhesion molecule is blocked, endothelial proliferation ceases and apoptosis ensues (Brooks et al., 1994; Friedlander et al., 1995).

5. Inhibition of Endothelial Cell Proliferation

We have seen that cell proliferation can be inhibited by removing factors that promote it. There is also evidence to show that there are endogenous molecules that directly inhibit endothelial cell proliferation. For example, interferon is able to block proliferation of bovine endothelial cells in culture, even in the presence of a positive stimuli such as bFGF (Heyns et al., 1985). Interferon was shown to inhibit early events of the G1 phase and the progression into S phase. Another factor reported to inhibit cell cycle progression of endothelial cells is SPARC (secreted protein, acidic and rich in cysteine), an extracellular, calcium-binding protein. This protein was shown to inhibit proliferation, delaying the entry of the cells into S phase (Funk and Sage, 1991; 1993). Platelet factor-4 has also been shown to inhibit endothelial cell proliferation and DNA synthesis, regardless of the positive stimuli present (Gupta and Singh, 1994). This peptide was shown to block entry of the cells into S phase and to turn off the cell cycle. Additional endogenous protein inhibitors of angiogenesis have been described, including factors derived from avascular cartilage (Tobelem, 1990). One new inhibitor that has been well characterized is angiostatin, a proteolytic fragment of plasmin derived from the action of a tumor cell line. This protein has been shown to block endothelial cell proliferation *in vitro*, as well as angiogenesis in animals (O'Reilly *et al.*, 1994).

Some factors that stimulate angiogenesis can also inhibit that process, usually when administered at higher concentrations. For example, studies suggest that IL-1 can inhibit endothelial cell growth *in vitro* and angiogenesis *in vivo* (Norioka *et al.*, 1987; Cozzoli *et al.*, 1990). This difference between this and its angiogenic stimulatory properties may be explained by dose and/or context-dependent activity. Another example is TNF-α; high doses can block endothelial cell proliferation. It remains to be seen whether these molecules have a bonafide role as angiogenesis inhibitors at high doses *in vivo*.

In addition to responding to proteins that can actively turn off proliferation, endothelial cells also respond to their physical environment. For example, they display a strong contact inhibition; cell growth is rapidly turned off when the endothelial cells have filled an area with a monolayer of cells. Because of this cell-cell contact and accompanying shape changes, a signal to turn off the cell cycle is conveyed. This signal appears to be important not only for cell cycle regulation, but also for blocking cells from entry into apoptosis (Re *et al.*, 1994). Yoshizumi *et al.* (1995) showed that cell-cell contact results in decreased levels of mRNA for cyclin A, and that this physical stimuli acts to transcriptionally regulate cell cycle factors.

6. Endothelial Cell Proliferation and Cell Cycle Events

The endothelial cell cycle can be restarted after long periods of quiescence, and the process of angiogenesis can then resume and continue. One of the best examples of this comes from the normal physiological process of corpus luteum formation, during which a carefully controlled proliferation of endothelial cells leads to production of new blood vessels, followed by regression and a period of inactivity (Gaede *et al.*, 1985).

In many cases, non-proliferating endothelial cells can be stimulated to grow again by adding a required growth factor. For example, G1-arrested bovine aortic endothelial cells were found to enter S phase and grow after addition of bFGF (Baldin *et al.*, 1990). Interestingly, this effect appeared to require uptake of bFGF and its translocation to the nucleus. This transport of bFGF to the nucleus and its effects on cell cycle were also seen in human umbilical vein endothelial cells (Imamura *et al.*, 1994). The effects of bFGF on endothelial cells were only seen in cells in late G1, suggesting a cell cycle-specific effect of the growth factor (Baldin *et al.*, 1990). The ability of growth factors such as bFGF, GM-CSF, PDGF and VEGF to restart endothelial cell proliferation may be key to their roles as promoters of angiogenic responses in pathological settings. In addition, these growth factors act on specific receptors, some of which (e.g., FLT-1, KDR, Tie) are specific for proliferating endothelial cells (Sato *et al.*, 1995; Fong *et al.*, 1995). The regulation of these receptors during the endothelial cell cycle remains to be studied.

The restart of endothelial cell proliferation is presumed to occur as a result of the upregulation of key regulatory proteins of the cell cycle. For example, there is an upregulation of cyclin E in endothelial cells found in breast cancer histological specimens (Dutta *et al.*, 1995). This area has been little studied in human endothelial cells. In other cells, mutations of cell cycle regulatory proteins have been associated with oncogenic events (Cordon-Cardo, 1995). Endothelial cell proliferation can continue uncontrolled after oncogenic transformation (Wagner and Risau, 1994). Expression of an oncogene in mouse endothelial cells results in the ability of injected cells to develop into hemangiomas, vascular tumors characterized by uncontrolled endothelial cell growth. However, the regulation of endothelial cell cycle regulatory proteins during the switch between proliferation and quiescence is not currently understood. It is also not clear whether there are any differences between endothelial cells and other cells in their cell cycle regulatory mechanisms.

7. Pharmacological Intervention in the Angiogenic Cell Cycle

As described above, endothelial cells are normally cell cycle arrested, but can reenter the cycle after the appropriate stimuli. When these stimuli arise from pathological sources, such as cancer or inflammation, it may be therapeutically desirable to block the reentry of endothelial cells into the cycle and thus obstruct the process of angiogenesis.

One compound that has been used for this effect is TNP-470, a compound derived from the natural product fumagillin, with proven angiogenesis inhibitory effects in animals (Kusaka *et al.*, 1991). In synchronized and stimulated human umbilical vein endothelial cells, TNP-470 inhibited DNA synthesis at concentrations that had no effect on cell viability (Kusaka *et al.*, 1994). In addition, TNP-470 has been shown to block the expression of cyclin D1 mRNA (Hori *et al.*, 1994). In another study, Antoine *et al.* (1994) showed that TNP-470 blocked entry of proliferating endothelial cells into the G1 phase of the cell cycle, a phase in which D cyclins play a role (Sherr, 1993). However, the compound failed to block transformed endothelial cells, suggesting that transformed cells can bypass the regulatory point in the cell cycle affected by TNP-470. Abe *et al.* (1994) showed that TNP-470 acted late in G1 and blocked the activation of both cdc2 and cdk2.

Drugs such as gold compounds and methotrexate have been used to treat rheumatoid arthritis for many years, and it has recently become clear that these compounds have anti-angiogenic activity which could contribute to their efficacy. Matsubara and Ziff (1987) studied the effects of gold sodium thiomalate, gold chloride, and auranofin on the proliferation of endothelial cells *in vitro*. Each effectively inhibited both basal and growth factor-induced ^{3}H-thymidine incorporation at concentrations comparable to plasma levels associated with clinically effective doses. Koch *et al.* (1988) also studied the effects of gold compounds *in vitro*, and demonstrated potent inhibition of angiogenesis factor secretion by macrophages without altering viability or protein synthesis, suggesting a second mechanism of angiogenesis inhibition for these compounds. Methotrexate was also shown to inhibit endothelial cell DNA synthesis *in vitro* at physiologically relevant concentrations (Hirata *et al.*, 1989). In addition, these investigators observed inhibition of angiogenesis by methotrexate *in vivo*, using an endothelial cell growth factor-induced corneal angiogenesis model.

Another compound that has been studied is sulfasalazine, which is used in chronic inflammatory diseases such as rheumatoid arthritis and inflammatory bowel disease. In proliferating bovine endothelial cells, sulfasalazine inhibited cell proliferation and the progression of cells through S phase (Sharon *et al.*, 1992). However, the mechanism of this effect remains to be uncovered. Genistein is a compound reported to inhibit the activity of receptor tyrosine kinases and thus block signal transduction of some growth factors. However, its effects on receptor tyrosine kinases may not completely explain its ability to inhibit angiogenesis, and a recent report suggests that its effects on the endothelial cell cycle may need closer inspection (Peterson, 1995).

It is clear that the many compounds that block endothelial cell proliferation may affect the endothelial cell cycle. Some of the challenges of this line of research are to examine these effects more closely, to determine whether compounds are directly affecting key components of the cell cycle apparatus, or whether the effects on cell cycle are secondary to the effects that compounds have on other aspects of cell function. Additional questions remain: Are there differences between the cell cycle mechanisms of endothelial cells and other cells? Are there differences in the cell cycles of endothelial cells proliferating in response to normal versus pathological conditions? The answers to these questions would further our understanding of endothelial cell function, and might also provide a potential drug target that can specifically affect cell cycle events in endothelial cells.

8. Summary

The process of angiogenesis is carefully orchestrated and controlled. One of the key controlling points is the decision of endothelial cells to arrest or to proceed through the cell cycle. Many factors can promote endothelial cell proliferation and removing these factors can halt it. Currently, the best approach to blocking angiogenesis appears to be the targeting of specific growth factors, such as VEGF. However, angiogenesis is so multifactorial that specific growth factors may not be involved in all pathological angiogenesis. There is a growing body of evidence that other factors can directly inhibit endothelial cell proliferation and promote cycle arrest, providing hope for a more universal approach to inhibition of angiogenesis. Some drugs can also affect the cell cycle status of endothelial cells and thus block proliferation and angiogenesis. However, more information is needed on the control points within the endothelial cell cycle in order to determine the ability of drugs to optimally and specifically block endothelial cell proliferation and the resulting angiogenesis.

References

Abe, J., Zhou, W., Takuwa, N., Taguchi, J., Kurokawa, K., Kumada, M., and Takuwa, Y., 1994, A fumagillin derivative angiogenesis inhibitor, AGM-1470, inhibits activation of cyclin-dependent kinases and phosphorylation of retinoblastoma gene product but not protein tyrosyl phosphorylation or protooncogene expression in vascular endothelial cells. *Cancer Res.*, **54**:3407–3412.

Antoine, N., Gremiers, R., De Roanne, C., Kusaka, M., Heinen, E., Simar, L.J., and Castronovo, V., 1994, AGM-1470, a potent angiogenesis inhibitor, prevents the entry of normal but not transformed endothelial cells into the G1 phase of the cell cycle. *Cancer Res.*, **54**:2073–2076.

Auerbach, W., and Auerbach, R., 1994, Angiogenesis inhibition: A review. *Pharmacol. Ther.*, **63**:265–311.

Baggiolini, M., and Clark-Lewis, I., 1992, Interleukin-8, a chemotactic and inflammatory cytokine. *FEBS Lett.*, **307**:97–101.

Baird, A., and Walicke, P.A., 1989, Fibroblast growth factors. *Br. Med. Bull.*, **45**:438–452.

Baldin, V., Roman, A.M., Bosc-Bierne, I., Amalric, F., and Bouche, G., 1990, Translocation of bFGF to the nucleus is G1 phase cell cycle specific in bovine aortic endothelial cells. *EMBO J.*, **9**:1511–1517.

Bell, L., and Madri, J.A., 1989, Effect of platelet factors on migration of cultured bovine aortic endothelial and smooth muscle cells. *Circ. Res.*, **65**:1057–1065.

Ben-Av, P., Crofford, L.J., Wilder, R.L., and Hla, T., 1995, Induction of vascular endothelial growth factor expression in synovial fibroblasts by prostaglandin E and interleukin-1: a potential mechanism for inflammatory angiogenesis. *FEBS Lett.*, **372**:83–87.

Benezra, D., 1978, Neuvasculogenic ability of prostaglandins, growth factors, and synthetic chemoattractants. *Am. J. Ophthal.*, **86**:455–461.

Benezra, D., Hemo, I., and Maftzir, G., 1990, *In vivo* angiogeneic activity of interleukins. *Arch. Opthamol.*, **108**:573–576.

Brooks, P.C., Montgomery, A.M., Rosenfeld, M., Reisfeld, R.A., Hu, T., Klier, G., and Cheresh, D.A., 1994, Integrin alpha v beta 3 antagonists promote tumor regression by inducing apoptosis of angiogenic blood vessels. *Cell*, **79**:1157–1164.

Bussolino, F., Wang, J.M., Defilippi, P., Turrini, F., Sanavio, F., Edgell, C.O.S., Aglietta, M., Arese, P., and Mantovani, A., 1989, Granulocyte- and granulocyte-macrophage-colony stimulating factors induce human endothelial cells to migrate and proliferate. *Nature*, **337**:471–473.

Bussolino, F., Ziche, M., Wang, J.M., Alessi, D., Morbidelli, L., Cremona, O., Bosia, A., Marchisio, P.C., and Mantovani, A., 1991, *In vitro* and *in vivo* activation of endothelial cells by colony-stimulating factors. *J. Clin. Invest.*, **87**:986–995.

Cohen, T., Nahari, D., Cerem, L.W., Neufeld, G., and Leviz, B.Z., 1996, Interleukin-6 induces the expression of vascular endothelial growth-factor. *J. Biol. Chem.*, **271**:736–741.

Colville-Nash, P.R., and Scott, D.L., 1992, Angiogenesis and rheumatoid arthritis: pathogeneic and therapeutic implications. *Ann. Rheum. Dis.*, **51**:919–925.

Colville-Nash, P.R., and Seed, M.P., 1993, The current state of angiostatic therapy, with special reference to rheumatoid arthritis. *Curr. Opin. Invest. Drugs*, **2**:763–813.

Cordon-Cado, C., 1995, Mutations of cell cycle regulators. Biological and clinical implications for human neoplasia. *Am J. Pathol.*, **147**:545–560.

Cozzoli, Torcia, M., Aldinucci, D., Ziche, M., Almerigogna, F., Bani, D., and Stern, D.M., 1990, Interleukin-1 is an autocrine regulator of endothelial cell growth. *Proc. Natl. Acad. Sci. USA*, **87**:6487–6491.

D'Amore, P., 1992, Mechanisms of endothelial growth control. *Am. J. Respir. Cell. Mol. Biol.*, **6**:1–8.

Dutta, A., Chandra, R., Leiter, L.M., and Lester, S. Cyclins as markers of tumor proliferation: immunocytochemical studies in breast cancer. *Proc. Natl. Acad. Sci. USA*, **92**:5386–5390.

Ernestein, J., Waleh, N.S., and Kramer, R.H., 1992, Basic FGF and TGF-beta differentially modulate integrin expression of human microvascular endothelial cells. *Exp. Cell Res.*, **203**:499–503.

Fajardo, L.F., Kwan, H.H., Kowalski, J., Prionas, S.D., and Allison, A.C., 1992, Dual role of tumor necrosis factor-a in angiogenesis. *Am. J. Pathol.*, **140**:539–544.

Folkman, J., 1995, Angiogenesis in cancer, vascular, rheumatoid and other disease. *Nature Medicine*, **1**:27–31.

Folkman, J., and Bream, H., 1992, Angiogenesis and inflammation. In *Inflammation: Basic Principles and Clinical Correlates, Second Edition*, ed. by J.I. Gallin, I.M. Goldstcin and R., Synderman, pp. 821–839, Raven Press, Ltd, New York.

Folkman, J., and Shing, Y., 1992, Angiogenesis. *J. Biol. Chem.*, **267**:10931–10934.

Fong, G.-H., Rossant, J., Gertsenstein, M., and Breltman, M.L., 1995, Role of the Flt-1 receptor tyrosine kinase in regulating the assembly of vascular endothelium. *Nature*, **376**:66–70.

Form, D.M., and Auerbach, R., 1983, PGE$_2$ and angiogenesis. *Proc. Soc. Exp. Biol. Med.*, **172**:214–218.

Friedlander, M., Shaffer, R., Kincaid, C., Brooks, P., and Cheresh, D., 1995, An antibody to the integrin α-v-β-3 inhibits ocular angiogenesis. *Invest. Ophthalmol. Visual Sci.*, **36**:S1047.

Funk, S.E., and Sage, E.H., 1991, The Ca2(+)-binding glycoprotein SPARC modulates cell cycle progression in bovine aortic endothelial cells. *Proc. Natl. Acad. Sci. USA*, **88**:2648–2652.

Funk, S.E., and Sage, E.H., 1993, Differential effects of SPARC and cationic SPARC peptides on DNA synthesis by endothelial cells and fibroblasts. *J. Cell. Physiol.*, **154**:53–63.

Gaede, S.D., Sholley, M.M., and Quattropani, S.L., 1985, Endothelial mitosis during the initial stages of corpus luteum neovascularization in the cycling adult rat. *Am. J. Anat.*, **172**:173–180.

Gupta, S.K., and Singh, J.P., 1994, Inhibition of endothelial cell proliferation by platelet factor-4 involves a unique action on S phase progression. *J. Cell. Biol.*, **127**:1121–1127.

Haralabopoulos, G.C., Grant, D.S., Kleinman, H.K., Lelkes, P.I., Papaioannou, S.P., and Maragoudakis, M.E., 1994, Inhibitors of basement membrane collagen synthesis prevent endothelial cell alignment in Matrigel in vitro and angiogenesis *in vivo*. *Lab. Invest.*, **71**:575–582.

Heyns, A.D., Eldor, A., Vlodavsky, I., Kaiser, N., Fridman, R., and Panet, A., 1985, The antiproliferative effect of interferon and the mitogenic activity of growth factors are independent cell cycle events. Studies with vascular smooth muscle cells and endothelial cells. *Exp. Cell. Res.*, **161**:297–306.

Hirata, S., Matsubara, T., Saura, R., Tateishi, H., and Hirohata, K., 1989, Inhibition of *in vitro* vascular endothelial cell proliferation and *in vivo* neovascularization by low-dose methotrexate. *Arthritis Rheum.*, **32**:1065–1073.

Hori, A., Ikeyama, S., and Sudo, K. Suppression of cyclin D1 mRNA expression by the angiogenesis inhibitor TNP-470 (AGM-1470) in vascular endothelial cells. *Biochem. Biophys. Res. Commun.*, **204**:1067–1073.

Imamura, T., Oka, S., Tanahashi, T., and Okita, Y., 1994, Cell cycle-dependent nuclear localization of exogenously added fibroblast growth factor-1 in BALB/c 3T3 and human vascular endothelia cells. *Exp. Cell Res.*, **215**:363–372.

Ingber, D.E., 1992, Extracellular matrix as a solid-state regulator in angiogenesis: identification of new targets for anti-cancer therapy. [Review]. *Seminars in Cancer Biol.*, **3**:57–63.

Koch, A.E., Cho, M., Burrows, J. Leibovich, S.J., and Polverini, P.J., 1988, Inhibition of production of macrophage-derived angiogeneic activity by the anti-rheumatic agents gold sodium thiomalate and auranofin. *Biochem. Biophys. Res. Commun.*, **154**:205–212.

Koch, A.E., Harlow, L.A., Haines, G.K., Amento, E.P., Unemori, E.N., Wong, W.L., Pope, R.M., and Ferrara, N., 1994, Vascular endothelial growth factor a cytokine modulating endothelial function in rheumatoid arthritis. *J. Immunol.*, **152**:4149–4156.

Kusaka, M., Sudo, K., Fujita, T., Marui, S., Itoh, F., Ingber, D., and Folkman, J., 1991, Potent anti-angiogenic action of AGM-1470: comparison to the fumagilliam parent. *Biochem. Biophys. Res. Commun.*, **174**:1070–1076.

Kusaka, M., Sudo, K., Matsutani, E., Kozai, Y., Marui, S., Fujita, T., Ingber, D., and Folkman, J., 1994, Cytostatic inhibition of endothelial cell growth by the angiogenesis inhibitor TNP-470 (AGM-1470). *Br. J. Cancer*, **69**:212–216.

Leibovich, S.J., Polverini, P.J., Shepard, H.M., Wiseman, D.M., Shivly, V., and Nuseir, N., 1987, Macrophage-induced angiogenesis is mediated by tumour necrosis factor-α. *Nature*, **329**:630–532.

Leung, D.W., Cachianes, G., Kuang, W.-J., Goeddel, D.V., and Ferrara, N., 1989, Vascular endothelial growth factor is a secreted angiogenic mitogen. *Science*, **246**:1306–1309.

Li, J., Perrella, M.A., Tsai, J.-C., Yet, S.-F., Hsieh, C.-M., Yoshizumi, M., Patterson, C., Endege, W.O., Zhou, F., and Lee, M.-E., 1995, Induction for vascular endothelial growth factor gene expression by interleukin-1b in rat aortic smooth muscle cells. *J. Biol. Chem.*, **270**:308–312.

Matsubara, T., and Ziff, M., 1987, Inhibition of human endothelial cell proliferation by gold compounds. *J. Clin. Invest.*, **79**:1440–1446.

Motro, B., Itin, A., Sachs, L., and Keshet, E., 1990, Pattern of interleukin 6 gene expression *in vivo* suggests a role for this cytokine in angiogenesis. *Proc. Natl. Acad. Sci. USA*, **87**:3092–3096.

Nagashima, M., Yoshino, S., Ishiwata, T., and Asano, G., 1995, Role of vascular endothelial growth factor in angiogenesis of rheumatoid arthritis. *J. Rheum.*, **22**:1624–1630.

Neufeld, G., Tessler, S., Gitay-Goren, H., Cohen, T., and Levi, B.-Z., 1994, Vascular endothelial growth factor and its receptors. *Prog. Growth Factor Res.*, **5**:89–97.

Norioka, K., Hara, M., Kitani, A., Hirose, T., Hirose, W., Harigai, M., Suzuki, K., Kawakami, M., Tabata, H. Kawagoe, M., and Nakamura, M., 1987, Inhibitory effect of human recombinant interleukin-1 alpha and beta on growth of human vascular endothelial cells. *Biochem. Biophys. Res. Commun.*, **145**:969–975.

O'Reilly, M.S. Holmgren, L., Shing, Y, Chem, C., Rosenthal, R.A., Moses, M., Lane, W.S., Cao, Y., Sage, E. H., and Folkman, J., 1994, Angiostatin: A novel angiogenesis inhibitor that mediates the suppression of metastases by a Lewis lung carcinoma. *Cell*, **79**:315–328.

Peterson, G., 1995, Evaluation of the biochemical targets of genistein in tumor cells. *J. Nutrition*, **125**:784S–789S.

Re, F., Zanetti, A., Sironi, M., Polentarutti, N., Lanfrancone, L., Dejana, E., and Colotta, F., 1994, Inhibition of anchorage-dependent cell spreading triggers apoptosis in cultured human endothelial cells. *J. Cell Biol.*, **127**:537–546.

Risau, W., 1994, Angiogenesis and endothelial cell function. *Arzneimittel-Forschung*, **44**:416–417.

Sato, T.N., Tozawa, Y., Deutsch, U., Wolburg-Buchholz, K., Fujiwara, Y., Gendron-Maguire, M., Gridley, T., Wolburg, H., Risau, W., and Qin, Y., 1995, Distinct roles of the receptor tyrosine kinases Tie1 and Tie2 in blood vessel formation. *Nature*, **376**:70–74.

Schreiber, A.B., Winkler, M.E., and Derynck, R., 1986, Transforming growth factor-a: a more potent angiogenic mediator than epidermal growth factor. *Science*, **232**:1250–1253.

4

Complex Regulation of the NF-κB Transcription Factor Complex: NF-κB activation is Inhibitable by Tyrosine Kinase Inhibitors

STEVEN M. TAFFET* and DALTON FOSTER

1. Introduction

As a consequence of infection by gram negative bacteria, an overly robust, and sometimes lethal activation of the host's immune response can occur (Morrison and Ryan, 1987). The chief component of gram negative bacteria responsible for driving such a powerful inflammatory response is bacterial lipopolysaccharide (LPS), which is an outer membrane glycolipid constituent of gram negative bacteria cell walls. This capacity of LPS to stimulate the host immune response is due, in part, to the fact that LPS is a potent activator of macrophages, which release a diverse arsenal of immunoregulatory molecules in response to contact with LPS (Watson et al., 1994). Some of the immune mediators secreted include proinflammatory cytokines such as IL-1α, IL-6, and tumor necrosis factor-α (Cavaillon and Cavaillon, 1990; Liberman and Baltimore, 1990; Drouet et al., 1990; Taffet et al., 1989). These immunological agents serve to both potentiate and propagate an immune response geared towards the eventual clearance of the bacterial infection. LPS-stimulated release of these mediators by macrophages is controlled mechanistically at many different levels. Although post-transcriptional, translational, and post-translation mechanisms all represent important levels of

*Correspondence: S.M. Taffet, The Department of Microbiology and Immunology, State University of New York Health Science Center at Syracuse, 750 East Adams Street, Syracuse, N.Y. 13210, USA. Tel: (315) 464-5419; Fax: (315) 464-4417; Taffets@VAX.CS.HSCSYR.EDU

control, the increase in cytokine gene transcription has been shown to be a primary site of regulation (Buetler et al., 1986).

The transcription factor NF-κB is ubiquitously expressed in nature. Nevertheless, the functioning of this transcription factor has been associated primarily with the transcriptional regulation of cellular and viral genes induced in response to both immune and inflammatory signals (Baeuerle and Henkel, 1994). Moreover, the induction of a variety of LPS-inducible genes in macrophages has been linked to the capacity of LPS to activate the transcription factor NF-κB (Müller et al., 1993).

NF-κB belongs to the NF-κB/Rel/Dorsal family of transcription factors. NF-κB/Rel/Dorsal family members are related by the fact that the genes which encode them contain a region of extensive homology among family members. This region, which is located within the N-terminal region of NF-κB/Rel/Dorsal proteins, is called the Rel homology domain (RHD). The RHD contains regions essential for the dimerization, DNA-binding, and nuclear localization of NF-κB/Rel/Dorsal family members (Miyamoto and Verma, 1995). Through the process of molecular cloning, many NF-κB related proteins have been identified. Some known members of the NF-κB/Rel/Dorsal family include the transcription factors p50 (NF-κB1) and its precursor p105 (Ghosh et al., 1990; Kieran et al., 1990); p65 (Rel A) (Nolan et al., 1990; Ruben et al., 1992); p52 (NF-κB2) and its precursor p98 (Mercurio et al., 1992); c-Rel (Inoue et al., 1991); and the related drosophila morphogen, dorsal (Ip et al., 1991). Both the p50 and p50B proteins are formed as a result of the proteolytic processing of the larger p105 and p98 proteins respectively (Mercurio et al., 1993; Palombella et al., 1994; Mellitis et al., 1993; Zheng et al., 1993). By virtue of the dimerization domains located within the RHD, these transcription factors are capable of homodimerizing or heterodimerizing, in order to form a variety of transcription factor complexes with varying DNA-binding specificities and transactivation potentials (Lernbecher et al., 1993; Perkins et al., 1992; Duckett et al., 1993).

Probably one of the more well studied transcription factors composed of members of the NF-κB/Rel/Dorsal family is NF-κB. NF-κB is formed as a result of the heterodimerization of the p65 and p50 proteins. Inactive p65/p105 complexes can also be formed, and subsequently processed, in activated cells to form p65/p50 complexes (Miyamoto and Verma, 1995). When cells are in an inactive state, NF-κB typically resides within the cytoplasm complexed with an ankyrin motif-rich inhibitory molecule called IκB (Baeuerle and Baltimore, 1988). IκB proteins are able to sequester NK-κB in the cytoplasm and inhibit DNA-binding by effectively masking both the nuclear localization and DNA binding domains found in NF-κB/Rel/Dorsal family members (Beg and Baldwin, 1993). There are a number of molecules known to function as IκB proteins. Members of the family of IκB proteins include the IkBα (Ghosh and Baltimore, 1990), IκB-β (Link et al., 1992; Thompson et al., 1995), IkB-γ (Inoue et al., 1992), p105 (Liou et al., 1992), p98, and Bcl-3 molecules (Franzoso et al., 1993; Hatada *et al.*, 1992). The stimulation of cells, with agents such as LPS, activates NF-κB by inducing the phosphorylation and subsequent proteolytic degradation of I-κB proteins (Lin *et al.*, 1995). The phosphorylation of I-κB proteins is a critical event in the process of NF-κB activation. Although the phosphorylation of IκB proteins alone cannot cause the dissociation of IκB from NF-κB (Didonato *et al.*, 1995; Irit *et al.*, 1995; Lin *et al.*, 1995), phosphorylation appears to target this molecule for ATP-dependent proteolysis (Lin *et al.*, 1995; Finco *et al.*, 1994; Traechner *et al.*, 1994), which leaves NF-κB free to nuclear translocate, bind to DNA, and activate the transcription of NF-κB-dependent genes (Baeuerle and Henkel, 1994).

The signal transduction pathway by which LPS activates NK-κB in mononuclear phagocytes has not been fully resolved, and is still the subject of considerable debate. The matter is complicated further by the fact that LPS appears to be promiscuous in its ability to activate protein kinase systems (Ulevitch and Tobias, 1994), an increasing number of which have been demonstrated to be capable of inactivating IκB proteins (Link *et al.*, 1992, Diaz-Meco *et al.*, 1994; Han *et al.*, 1994). The stimulation of macrophages by LPS in particular can activate protein kinases such the Raf-1 protein kinase, mitogen-activated protein kinases (MAPK), both classical and non-classical protein kinase C isozymes, and cyclic AMP-dependent protein kinases (PKA) as well as protein tyrosine kinases (Hambleton *et al.*, 1995; Büscher *et al.*, 1995; Geng *et al.*, 1993; Lui *et al.*, 1994; Weinstein *et al.*, 1992; Dong *et al.*, 1993). We have previously conducted experiments in our laboratory using RAW264.7 murine macrophage-like cells in order to determine the relevance of protein kinase C as well as PKA in the process of NF-κB activation in response to LPS. From those studies it was determined that the LPS signal transduction pathway does not require the activation of classical PKCs, as neither the down regulation of PKC by phorbol esters, nor pretreatment with the PKC inhibitor H-7, prevented the activation of NF-κB by LPS. Additionally, it was concluded that the activation of PKA in RAW264.7 cells was not critical for the activation of NF-κB, as treatment of RAW264.7 cells with PKA agonists could not induce NF-κB activation (Vincenti *et al.*, 1992). Further clarifying this issue, Hambleton (1995) recently demonstrated that the activation of NF-κB by LPS in RAW264.7 cells occurs independently of LPS-induced activation of MAP and RAF-1 kinases, as the LPS-independent activation of the protein kinases did not lead to the activation of NF-κB.

There has been, however, an accumulation of evidence suggesting the involvement of tyrosine kinases in the process by which LPS stimulates NF-κB activation in mononuclear phagocytes. This evidence was based primarily upon the ability of tyrosine kinase inhibitors to inhibit a number of LPS-mediated responses in monocytes and macrophages, as well as their ability to inhibit the activation of NF-κB in a number of cell types tested. For example, the LPS-induced transcription of the IL-1, IL-6, and TNF-α cytokine genes, for which NF-κB activation is critical, has been demonstrated to be sensitive to inhibition by the tyrosine kinase inhibitors herbimycin A and genistein (Geng *et al.*, 1993;

Novogrodsky *et al.*, 1994). Tyrosine kinase inhibitors have also been shown to inhibit the LPS-stimulated tyrosine phosphorylation and activation of MAP kinases in both human monocytes and murine macrophages (Weinstein *et al.*, 1992; Dong *et al.*, 1993). Moreover, herbimycin A has recently been demonstrated to inhibit the LPS-mediated activation of the tyrosine kinases $p53/56^{lyn}$, $p58^{hck}$, and $p59^{c-fgr}$ (Štefanová *et al.*, 1991, 1993) in human monocytes.

In order to determine whether tyrosine kinase activity might be involved in the activation of NF-κB by LPS in the RAW264.7 murine macrophage-like cell line, we utilized the tyrosine kinase inhibitors genistein, herbimycin A and tyrphostin. Pretreatment of RAW264.7 cells with any of the inhibitors resulted in an inhibition of the LPS-induced activation of NF-κB in RAW264.7 macrophage-like cells. Additionally, we have demonstrated that the LPS-induced processing of the p105 protein was also sensitive to inhibition by tyrosine kinases inhibitors. These results indicate that LPS activation of NF-κB and processing of p105 in RAW264.7 occur via a tyrosine kinase-dependent pathway.

2. Materials and Methods

2.1. Cell Culture and Stimulation

RAW 264.7 cells, an Abelson virus transformed murine macrophage-like cell line, were cultured in 100 mm² petri dishes with Dulbecco's modification of Eagle's Medium supplemented with 15mM HEPES, penicillin-streptomycin, and 5% supplemented calf serum (Hyclone, Logan UT). Cells were maintained at 37°C in 5% CO_2 humidified air. For the inhibition studies RAW264.7 cells were pretreated with either genistein, tyrphostin or herbimycin A (GIBCO-BRL Life Technologies, Grand Island, MD) for 2 hr prior to stimulation with 100 ng/ml of chromatographically purified LPS 0111:B4 (Sigma, St. Louis MO) for 1 hr. All tissue culture reagents and inhibitors were determined to be endotoxin-free by E-toxate assay at a level of sensitivity of 0.1 ng/ml (Sigma, St Louis MO).

2.2. Preparation of Nuclear and Cytoplasmic Extracts

Nuclear extracts were prepared as described by Vincenti (1992). Briefly, cells were washed twice in Tris buffered saline, pH 7.9, prior to resuspension in 400 [μ]l of ice cold buffer A (10 mM HEPES, pH 7.9; 10 mM KC1; 0.1 mM EDTA; 0.1 mM EGTA; 1mM DTT; 0.5 mM PMSF; 1 μM aprotinin; 14 μM leupeptin; 80 μg/ml benzamidine; 1 μM pepstatin). After allowing cells to swell on ice for 15 min, cell lysis was achieved by the addition of NP-40 (final concentration 0.6% V/V) followed by vigorous vortexing for 10 sec. The nuclei were then separated from the cytosol by centrifugation at 13, 000× G for 30 sec, with the resulting nuclear pellet being resuspended in 50 μl ice cold buffer C (20 mM HEPES, pH 7.9; 0.4 M NaCl;

1 mM EDTA; 1 mM EGTA; 1 mM DTT; 1 mM PMSF). Following agitation for 15 min at 4°C by Vortex Genie (Scientific Industries, Inc., Bohemia, NY), nuclear extracts were centrifuged at $13,000\times$ G for 5 min, aliquoted, and subsequently stored at $-70°$C. Protein concentrations were determined by Bradford Assay (Bradford, 1976).

2.3. Electrophoretic Mobility Shift Assay (EMSA)

EMSA was performed as previously described by Vincenti (1992). The probes used to detect NF-κB activity consisted of a 27-mer containing the -510 κB site from the murine TNF-α promoter.

5′AGCTCAAACAGGGGGCTTTCCCTCCTC   3′
3′   GTTTGTCCCCCGAAAGGGAGGAGTCGA 5′

The Tumor Necrosis Factor-α intronic enhancer probe (TNFiENH), used as a control, consisted of the following sequence:

5′–GATCCAGAGGGTGCAGGAACCGGAAGTG–3′
3′–GTCTCCCACGTCCTTGGCCTTCACTCGA–5′

2.4. Western Blotting

Western blots were performed as previously described by Zheng *et al.* (1993). Briefly, 10 μg of nuclear protein was separated by 8% SDS-PAGE, and transferred to 0.45 μM nitrocellulose (Shleicher and Schuell) in Towbin transfer buffer (Towbin *et al.*, 1979) containing 10% methanol at 100 V for 1 hr. Blots were then processed as described in detail by Zheng (1993) using rabbit antisera purified against an NF-κB1 region (aa$^{350-500}$) lacking homology to other NF-κB/Rel family members. After exposure to a peroxidase conjugated goat anti-rabbit sera the immunoblot was visualized using enhanced chemiluminescence (Amersham, Arlington Heights IL).

3. Results

3.1. Concentration Dependent Inhibition of LPS-induced Activation of NF-κB DNA-binding Activity in RAW264.7 Macrophages by Herbimycin A

We have previously determined that LPS induction of NF-κB DNA binding activity in RAW264.7 murine macrophage-like cells occurs independent of

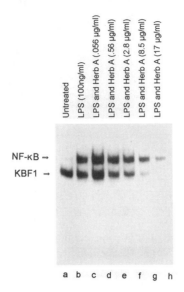

Figure 1. Herbimycin A inhibition of LPS-induced activation of NF-κB and KBF1 in RAW264.7 cells. Nuclear extracts were prepared from RAW264.7 cells, either untreated (lane a), LPS-treated (lane b), or from cells treated with increasing doses of herbimycin A for 2 hr prior to stimulation with 100 ng/ml LPS for 1 hr (lanes c-h). Nuclear extracts were analyzed for NF-κB binding activity by EMSA. The EMSA reaction mixture contained 30,000 cpm oligonucleotide probe and 5 μg nuclear extract. The radio-labeled oligoncleotide used as a probe contained the −510 kB-element from the murine TNF-α gene. The positions of NF-κB and KBF1 are marked.

classical PKC activation. To investigate whether the activation of tyrosine kinases may be involved in the process, the tyrosine kinase inhibitor herbimycin A was used. To do so, RAW264.7 cells were pretreated with herbimycin A prior to LPS stimulation, after which nuclear extracts were prepared and inhibition of LPS-induced NF-κB DNA-binding activity determined by EMSA analysis. As shown in Fig. 1, lane b, a detectable level of KBF1 (NF-κB1 homodimer) was present in unstimulated macrophages; however, there was no detectable NF-κB. Stimulation of RAW264.7 cells with 100 ng/ml LPS for 1 hr resulted in a significant increase in NF-κB levels, indicated by the increase in κB-element binding. Interestingly, pretreatment of the cells with a low concentration of herbimycin A (0.056 μg/ml) resulted in an enhancement of LPS-stimulated binding to the κB element (Fig. 1, lane c); however, subsequent increases in herbimycin A concentrations (2.8 μg/ml, 8.5 μg/ml, and 17 μg/ml) led to the dose-dependent decrease in the amount of κB-element binding induced by LPS. The inhibitory effect of herbimycin A was not strictly limited to NF-κB binding activity, as increases in KBF1 binding was also inhibited.

We have previously proposed that increases in KBF1 binding in response to LPS stimulation occurred as a result of processing of the p105 precursor (Zheng *et al.*, 1993). These results indicate that the formation of KBF1 as a result of processing of the p105 precursor and the release of NF-κB from IκB might both be tyrosine kinase dependent processes. To determine whether the inhibition of LPS-induced κB-element binding by herbimycin A at 17 μg/ml was due to toxicity or interference with nuclear integrity, nuclear extracts from untreated, LPS (100 ng/ml) treated, herbimycin A (17 μg/ml) treated, and herbimycin A (17 μg/ml)/LPS treated were prepared, and EMSA analysis was performed using both a κB specific probe, and the TNF-α intronic enhancer probe (TNFiENH). The TNFiENH probe contains a binding site for the *ets*-related transcription factor GABP (Foster *et al.*, in preparation), which binds constitutively to the TNFiENH in RAW264.7 cells. Consistent with the results shown in Fig. 1, LPS stimulation of RAW264.7 cells led to the significant induction of both NF-κB and KBF1 binding to the −510 κB probe (Fig. 2A, lane b), and once again pretreatment of cells with 17 μg/ml herbimycin A led to a significant decrease in both the KBF1 and NF-κB DNA binding activity induced by LPS (Fig. 2A, lane d). On the other hand control EMSAs performed using the same nuclear extracts with the intronic TNF-α enhancer probe showed very little variation in GABP binding activity (Fig. 2B). This indicated that the inhibition of NF-κB DNA-binding detected was due to a specific inhibition of LPS-induced NF-κB binding activity in RAW264.7 cells by herbimycin A, and not to a loss of nuclear integrity caused by the inhibitor.

3.2. Concentration Dependent Inhibition of LPS Induced Activation of NF-κB in RAW264.7 Macrophages by Tyrphostin and Genistein

To determine whether another tyrosine kinase inhibitor, tyrphostin, could inhibit LPS induced κB-element binding in a manner similar to herbimycin A, RAW264.7 macrophage-like cells were also pretreated with tyrphostin at increasing concentrations prior to stimulation by LPS. Fig. 3 demonstrates that tyrphostin at 0.5 μM was capable of significant inhibition of LPS-induced NF-κB and KBF1 binding. Fig. 3, lanes e-h, shows a concentration-dependent decrease in induced κB-element binding as the tyrphostin concentration was increased up to 2.5 μM. Treatment of RAW264.7 cells with 2.5 μM tyrphostin resulted in maximal inhibition of LPS-induced κB binding activity, as no further inhibition occurred when cells were treated with 5 μM tyrphostin. Control EMSA determined that there was no significant inhibition of TNFiENH binding activity due to tyrphostin treatment (data not shown). In addition to tyrphostin, a third tyrosine kinase inhibitor genistein also used. Fig. 4 demonstrates that, as with tyrphostin and herbimycin A, genistein was capable of causing significant inhibition of LPS-induced activation of NF-κB activation.

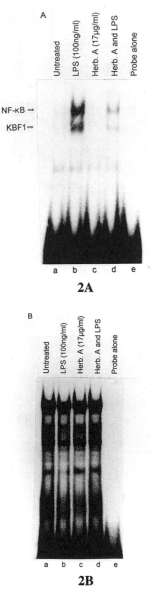

Figure 2. Herbimycin A treatment of RAW264.7 cells has no effect on GABP binding. Nuclear extracts were prepared from RAW264.7 cells, either untreated (lane a), LPS-treated (lane b), or from cells treated with 17 μg/ml herbimycin A for 2 hr prior to stimulation with 100 ng/ml LPS for 1 hr (lanes c, d). Nuclear extracts were analyzed by EMSA. The EMSA reaction mixture contained 30,000 cpm oligonucleotide probe and 5 μg nuclear extract. Panel A; The radio-labeled oligonucleotide used as a probe contained the −510 κB-element from the murine TNF-α gene. The positions of NF-κB and KBF1 are marked. Panel B; The probe used was the GABP binding site from the murine TNF-α gene.

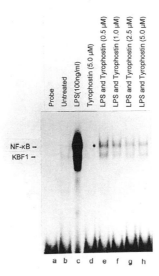

Probe
Untreated
LPS(100ng/ml)
Tyrophostin (5.0 µM)
LPS and Tyrophostin (0.5 µM)
LPS and Tyrophostin (1.0 µM)
LPS and Tyrophostin (2.5 µM)
LPS and Tyrophostin (5.0 µM)

NF-κB →
KBF1 →

a b c d e f g h

Figure 3. Tyrphostin inhibition of LPS-induced activation of NF-κB and KBF1 in RAW264.7 cells. Nuclear extracts were prepared from RAW264.7 cells that were untreated (lane b), LPS-treated (lane c), treated with tyrphostin alone (5.0 µM; lane d), or from cells treated with increasing doses of tyrphostin for 2 hr prior to stimulation with 100 ng/ml LPS for 1 hr (lanes e-f). Nuclear extracts were assayed for NF-κB binding by EMSA.

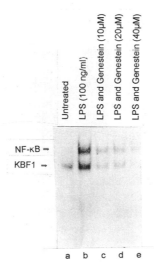

Untreated
LPS (100 ng/ml)
LPS and Genestein (10µM)
LPS and Genestein (20µM)
LPS and Genestein (40µM)

NF-κB →
KBF1 →

a b c d e

Figure 4. Genistein inhibition of LPS-induced activation of NF-κB and KBF1 in RAW264.7 cells. Nuclear extracts were prepared from RAW264.7 cells that were untreated (lane a), LPS-treated (lane b), or treated with increasing doses of genistein for 2 hr prior to stimulation with 100 ng/ml LPS for 1 hr (lanes c-e). Nuclear extracts were assayed for NF-κB binding by EMSA.

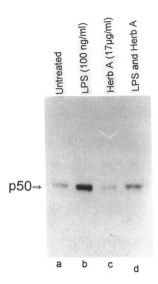

Figure 5. Herbimycin A inhibits LPS-induced NF-κB and KBF1 binding by preventing the LPS-induced nuclear translocation of the transcription factors. Nuclear extracts (the same extracts used in Fig. 2) were prepared from RAW264.7 cells, either untreated (lane a), LPS-treated (lane b), or from cells treated with 17 μg/ml herbimycin A for 2 hr prior to stimulation with 100 ng/ml LPS for 1 hr (lanes c,d). Western blot analysis was performed using affinity purified anti-p50 antiserum (1:700). The position of p50 is marked.

3.3. Herbimycin A Inhibition of LPS Induced κB-element binding is Due to Inhibition of LPS-induced Nuclear Localization of NF-κB and KBF1

Western immunoblot analysis was performed to determine whether inhibition of NF-κB and KBF1 binding activity by herbimycin A occurred due to inhibition of LPS-induced activation and subsequent nuclear localization of the transcription factors as opposed to a direct inhibition of DNA binding. To do so, nuclear extracts from untreated, LPS stimulated, herbimycin A treated, and herbimycin A pretreated/LPS stimulated RAW264.7 macrophages were probed using antisera generated against a region unique to the p50 (NF-κB1) subunit of NF-κB. Fig. 5, lane a, shows that very little p50 resides within the nucleus of unstimulated cells; however, consistent with EMSA data, LPS stimulation results in a dramatic increase in levels of nuclear p50 (Fig. 5, lane b). Fig. 5, lane d, shows that herbimycin A significantly inhibits LPS-induced nuclear localization of p50. This data indicates that the herbimycin A inhibition of LPS-induced κB binding activity detected by EMSA was mediated through the inhibition of induced nuclear localization of κB-element binding proteins.

3.4. Genistein Inhibits the LPS Induced Processing of the p105 Precursor

In a previous report, we have shown that LPS stimulation of RAW264.7 cells leads to the processing of the p105 protein into its p50 counterpart (Zheng *et al.*, 1993). Recent data has revealed that the processing of the p105 proteins, like proteolysis of IκB, occurs as a part of a stimulant induced, ATP-dependent, ubiquitin mediated process (Lin *et al.*, 1995; Finco *et al.*, 1994; Traeckner *et al.*, 1994). We were interested in determining whether tyrosine kinase activation was involved in the LPS-induced processing of p105 in RAW264.7 cells. Western blot analysis was performed on cytosolic extracts from RAW264.7 cells to determine whether the tyrosine kinase inhibitor genistein could inhibit the processing of p105 in response to LPS. As shown in Fig. 6, lane b, LPS stimulation of RAW264.7 cells caused a reduction in the amount of p105 found in the cytoplasm of RAW264.7 cells. A reduction in the amount of cytoplasmic p50 was also seen, resulting from the LPS-stimulated nuclear localization of the p50 containing complexes. Treatment of RAW264.7 cells with genistein alone had no effect on the cytoplasmic levels of p105 or p50; however, genistein was able to inhibit the LPS-stimulated processing of p105 and nuclear localization of p50.

4. Discussion

Activation of the transcription factor NF-κB is a crucial event required for the LPS-induced transcription of several immune and inflammatory response genes (Müller *et al.*, 1993). Although our broad understanding of the events required for the activation of this transcription factor are reasonably clear, the precise mechanisms involved in the process continue to elude us. For example, it is well known that the activation of NF-κB occurs as result of a signal induced release of NF-κB from cytoplasmic sequestration by IκB or as a result of processing of the p105 component of a p105/p65 complex (Miyamoto and Verma, 1995). Both mechanisms require an initial phosphorylation event which in and of itself, is unable to cause either dissociation of the transcription factor from IkB or processing of p105 (DiDonato *et al.*, 1995; Irit *et al.*, 1995). Instead, phosphorylation appears to target these proteins for proteolytic degradation by an ATP-dependent ubiquitin mediated process (Palonebella *et al.*, 1994). The phosphorylation of specific serine residues has been shown to be required for the site-specific signal-induced phosphorylation and proteolysis of the human IκB protein (Brown *et al.*, 1995). It is expected that residues of similar importance exist within the p105 polypeptide. What remains unclear about the process of LPS-induced activation of NF-κB and processing of p105, however, are the incidents leading up to the requisite phosphorylation event. Mapping of the signal transduction pathway resulting in the phosphorylation of the IκB and p105 proteins has been a difficult process, complicated by the fact that LPS stimulation

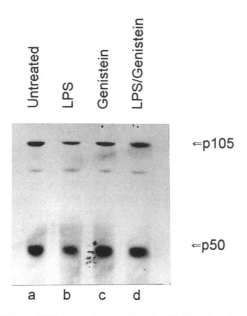

Figure 6. Genistein inhibition of LPS-induced processing of p105. Cytoplasmic extracts were prepared from RAW264.7 cells that were untreated (lane a), LPS-treated (lane b), treated with genistein alone (lane c), or treated with genistein for 2 hr prior to stimulation with 100 ng/ml LPS for 1 hr (lane d). Western blot analysis was performed using affinity purified anti-p50 antiserum (1:700). The positions of p105 and p50 are marked.

of macrophages activates a variety of serine-threonine specific protein kinases, a number of which have been demonstrated to be capable of phosphorylating IκB *in vitro* (Link *et al.*, 1992, Diaz-Meco *et al.*, 1994; Han *et al.*, 1994). Protein kinases activated by LPS include the family of classical and non-classical protein kinase C enzymes, MAP kinases, the cAMP-regulated protein kinase, the Raf-1 kinase, and the src-family of protein tyrosine kinases (Hambleton *et al.*, 1995; Büscher *et al.*, 1995; Geng *et al.*, 1993; Lui *et al.*, 1994; Weinstein *et al.*, 1992; Dong *et al.*, 1993). In a previous report we had attempted to shed some light on the issue of LPS-induced activation of NF-κB the RAW264.7 macrophage-like cell line. Based upon our studies, we concluded that the LPS-induced activation of NF-κB in RAW264.7 macrophage-like cells occurs independently of classical PKC and cAMP regulated kinase activity (Vincenti *et al.*, 1992).

Overall, there is significant support for a role for tyrosine phosphorylation in LPS induction of macrophage activity. Several reports demonstrate that LPS stimulates tyrosine phosphorylation and activation of MAP kinases in macrophages and macrophage-like cell lines (Weinstein *et al.*, 1992; Dong *et al.*, 1993). LPS-stimulated transcription and secretion of IL-1, IL-6, and TNF-α (Geng *et al.*, 1993) are blocked by inhibitors of tyrosine kinases, as is LPS-stimulated macrophage cytotoxicity (Dong *et al.*, 1993). More significantly, the activation of NF-κB by LPS in blood mononuclear phagocytes was shown to be blocked by genistein and herbimycin A (Geng *et al.*, 1993). The mechanism by which LPS stimulates tyrosine kinase activity has been addressed in studies demonstrating that the LPS receptor, CD14, co-immunoprecipitates with the *src* related tyrosine kinases $p53/p56^{lyn}$, 58^{hck}, and $p59^{c-fgr}$ (Štefanová *et al.*, 1991; 1993). Additionally, the LPS activation of the $p53/56^{lyn}$, 58^{hck}, and $p59^{c-fgr}$ kinases in human monocytes was inhibited by the tyrosine kinase inhibitor herbimycin A (Štefanová *et al.*, 1993). It was, therefore, important to determine whether LPS-induced activation of NF-κB occurred via a tyrosine kinase-dependent process.

We now report that the LPS-induced activation of NF-κB, as well as a related transcription factor KBF1, were inhibited by the tyrosine kinase inhibitors herbimycin A, genistein, and tyrophostin. Additionally, we demonstrated that another process stimulated by LPS, the processing of p105 was inhibited by the tyrosine kinase inhibitor genistein. Our results are in apparent conflict with a recent report by Delude *et al.* (1994) in which they were unable to inhibit activation of NF-κB by LPS in RAW264.7 cell using tyrophostin and genistein. This discrepancy may be due to their use of ineffective concentrations of these tyrosine kinase inhibitors. For example, Delude *et al.* (1994) unsuccessfully used 5 μg/ml herbimycin A to inhibit NF-κB activation, which, based upon the data in this manuscript, is too dilute to cause significant inhibition of LPS-induced DNA binding by NF-κB in RAW264.7 cells. Significant inhibition of both LPS-induced NF-κB and KBF1 binding was not seen until a concentration of 8.5 μg/ml herbimycin A was used. When the tyrosine kinase inhibitor genistein was used at 100 μM, Delude *et al.* (1994) again failed to detect inhibition of LPS induced activation of NF-κB in RAW264.7 cells. It is not clear why this concentration was ineffective at inhibiting NF-κB activation, as 100 μM genistein was far higher than the maximal concentration (40 μM genistein) employed in the studies using this inhibitor presented within this manuscript.

In addition to inhibition of LPS-induced transcription factor DNA binding activity, genistein was also able to inhibit the LPS-induced processing of the p105 protein. This indicates that the activation of tyrosine kinases plays a role in both the activation of NF-κB and the processing of p105. It is unlikely, however, that tyrosine kinases play a direct role in either process by targeting IκB and p105 for proteolysis. Although human IκB does contain a consensus site for phosphorylation by tyrosine kinases, mutation of this region could not prevent the signal induced phosphorylation and degradation of the protein (Brown *et al.*, 1995). Additionally, although signal-induced phosphorylation of p105 has been described by Neumann (1992), proteolysis has not been shown to occur as a result of phosphorylation. It appears that tyrosine phosphorylation of IkB or p105 is not directly responsible for activation. However, the data presented herein, as well as by others, indicates that the activation of tyrosine kinases may be both an early and a critical step in LPS-induced activation of NF-κB and KBF1.

5. Summary

The stimulation of macrophages by bacterial lipopolysaccharide induces the activation of the transcription factor NF-κB. We have previously demonstrated that this process in RAW264.7 murine macrophage-like cells is independent of activation of protein kinase C. Recent evidence has shown that treatment of macrophages with inhibitors of tyrosine kinases inhibits a variety of LPS-stimulated responses, including TNF-α transcription, MAP kinase activation, and the activation of the tyrosine kinases p53/56^{lyn}, p58^{hck}, and p59$^{c\text{-}fgr}$. We chose to determine whether LPS-activation of NF-κB in RAW264.7 cells could be similarly inhibited by tyrosine kinase inhibitors. Here we report that pretreatment of RAW264.7 cells with either herbimycin A, genistein, or tyrphostin significantly reduced NF-κB activation by LPS, as determined by electrophoretic mobility shift assay. This decrease correlated with an inhibition of the LPS-stimulated nuclear localization of the p50 (NF-κB1) subunit of NF-κB and processing of the p105 precursor as detected by western blot analysis. These results indicate that activation of NF-κB by LPS in RAW264.7 macrophage-like cells may be a tyrosine kinase-dependent process.

References

Baeuerle, P.A., and Baltimore, D., 1988, Activation of DNA binding in an apparently cytoplasmic precursor of the NF-κB transcription factor, *Cell*, **53**:211–217.

Baeuerle, P.A., and Henkel, I., 1994, Function and activation of NF-κB in the immune syste, *Annu. Rev. Immunol.*, **12**:141–179.

Beg, A.A., and Baldwin, A.S., Jr., 1993, The IkB proteins: multifunctional regulators of the Rel/NF-κB transcription factors, *Genes and Development*, 7:2064–2070.

Beutler, B., Krochin, N., Milsark, I.W., Leudke, C., and Cerami, A., 1986, Control of cachectin (tumor necrosis factor) synthesis: Mechanisms of endotoxin resistance, *Science*, **232**:977–980.

Bradford, M.M., 1976, A rapid and sensitive method for the quantition of microgram quantities of protein utilizing the principle of protein-dye binding, *Anal. Biochem.*, **72**:248–254.

Brown, K., Gerstberger, S., Carlson, L., Franzoso, G., and Siebelist, U., 1995, Control of IκB proteolysis by site-specific, signal-induced phosphorylation, *Science*, 267:1485–1488.

Büscher, D., Hipskind, R.A., Krautwald, S., Reimann, T., and Baccarini, M., 1995, Ras-dependent, and -independent pathways target the mitogen-activated protein kinase network in macrophages, *Mol. Cell. Biol.*, **15**:466–475.

Cavaillon, J.M., and Haeffner-Cavaillon, 1990, Signals involved in interleukin 1 synthesis and release by lipopolyssacharide stimulated monocytes/macrophages, *Cytokine*, **2**:313–329.

Delude, R.L., Fenton, M.J., Savedra, R., Jr., Perera, P.Y., Vogel, S.N., Thieringer, R., and Golenbock, D.T., 1994, CD14-mediated translocation of nuclear factor-κ B induced by lipopolysaccharide does not require tyrosine kinase activity, *J. Biol. Chem.*, **269**(35):22253–22260.

Diaz-Meco, M.T., Dominguez, I., Sanz, L., Dent, P., Lozano, J., Municio, M.M., Berra, E., Hay, R.T., Sturgill, T.W., and Moscat, J., 1994, ζPKC induces phosphorylation and inactivation of Iκb-α, *EMBO J.*, **13**(12):2842–2848.

DiDonato, J., Mercurio, F., and Karin, M., 1995, Phosphorylation of the IkBα precedes but is not sufficeint for its dissociation from NF-κB, *MOl. Cell. Biol.*, **15**:1302–1311.

Dong, Z., O'Brian, C., and Fidler, I., 1993, Activation of tumoricidal properties in macrophages by lipopolysacchardies requires protein-tyrosine kinase activity, *J. Leukoc. Biol.*, **53**:53–60.

Drouet, C., Shakhov, A.N., and Jongeneel, C.V., 1990, Enhancers and the transcription factors controlling the inducibility of the tumor necrosis factor-α promoter in primary macrophages, *J. Immunol.*, **147**:1694–1700.

Duckett, C.S., Perkins, N.D., Kowalkik, T.F., Schmid, R.M., Huang, Eng-Shuang, Baldwin, A.S., and Nabel, G.J., 1993, Dimerization of NF-κB2 with RelA (p65) regulates DNA binding, transcriptional activation, and inhibition by and IκB-α (MAD-3), *Mol. Cell. Biol.*, **13**:1315–1322.

Fan, C-M., and Maniatis, T., 1991, Generation of p50 subunit of NF-κB by processing of p105 through an ATP-dependent pathway, *Nature*, **354**:395–398.

Finco, T.S., Beg, A.A., and Baldwin, A.S., Jr., 1994, Inducible phosphorylation of IκBα is not sufficent for its dissociation from NF-κB and is inhibited by protease inhibitors, *Proc. Natl. Acad. Sci. USA*, **91**:11884–11888.

Franzoso, G., Bours, V., Park, P., Tomita-Yamaguchi, M., Kelly, K., and Siebenlist, U., 1992, The candidate oncoprotein Bcl-3 is an antagonist of p50/NF-κB inhibition, *Nature*, **359**:339–342.

Franzoso, G., Bours, V., Park, P., Tomita-Yamaguchi, M., Tomohiko, K., Brown, K., and Siebenlist, U., 1993, The oncoprotein Bcl-3 can facilitate NF-κB transactivation by removing inhibiting p50 homodimers from select κB sites. *EMBO J.*, **12**:3893–3901.

Geng, Y., Zhang, B., and Lotz, M., 1993, Protein tyrosine kinase activation is required for lipopolysac-charide induction of cytokines in human blood monocytes, *J. Immunol.*, **151**:6692–6700.

Ghosh, S., and Baltimore, D., 1990, Activation *in vitro*, of NF-κB by phosphorylation of its inhibitor IkB, *Nature*, **344**:1019–1029.

Ghosh, S.A., Gifford, A., Riviere, L., Tempst, P., Nolan, G.P., and Baltimore, D., 1990, Cloning of the p50 DNA-binding subunit of NF-κB:homology to rel and dorsal, *Cell*, **62**:1019–1029.

Hambleton, J., McMahon, M., and Defranco, A., 1995, Activation of RAf-1 and mitogen-activated protein kinase in murine macrophages partially mimics lipopolysaccharide-induced signaling events, *J. Exp. Med.*, **182**:147–154.

Han, J., Lee, J.D., Bibbs, L., and Ulevitch, R.J., 1994, A MAP kinase targeted by endotoxin and hyperosmolarity in mammalian cells, *Science*, **265**:808–811.

Hatada, E.N., Nieters, A., Wulczyn, F.G., Naumann, M., Meyer, R., Nucifora, G., McKeithan, T.W., and Scheidereit, C., 1992, The ankyrin repeat domains of the NF-κB precursor p105 and the protooncogene *bcl-3*, act as specific inhibitors of NF-κB DNA binding, *Pro. Acad. Natl. Sci. USA*, **89**:2489–2493.

Inoue, J., Kerr, D., Ransone, L.J., Bengal, E., Hunter, T., and Verma, I.M., 1991, *c-rel*, activates but *v-rel*, suppresses transcription from κB sites, *Proc. Natl. Acad. Sci. USA*, **88**:3715–3719.

Inoue, J., Kerr, L.D., Kakizuka, A., and Verma, I.M., 1992, IκB-γ, a 70 kDa protein identical to the c-terminal half of p110 NF-κB: a new member of the IκB family, *Cell*, **68**:1109–1120.

Ip, Y.T., Kraut, R., Levine, M., and Rushlow, C., 1991, The dorsal morphogen is a sequence-specific DNA-binding protein that interacts with a long range repression element in Drosophila, *Cell*, **64**:439–446.

Irit, A., Yaron, A., Hatzubai, A., Jung, S., Avraham, A., Gerlitz, O., Pashutlavon, I., and Ben-neriah, Y., 1995, *In vivo*, stimulation of IκB phosphorylation is not sufficient to activate NF-κB, *Mol. Cell. Biol.*, **15**:1294–1301.

Kieran, M., Blank, V., Logeat, F., Vandekerckhove, J., Lottspeich, F., Le Bail, O., Urban, M., Kourilsky, P., Baeuerle, P., and Israel, A., 1990, The DNA-binding subunit of NF-κB is identical to factor KBF1 and homologous to the rel oncogene product, *Cell*, **62**:1007–1018.

Lernbecher, T., Müller, U., Wirth, T., 1993, Distinct NF-κB transcription factors are responsible for tissue-specific and inducible gene activation, *Nature*, 365–770.

Li, S., and Sedivy, J.M., 1993, Raf-1 protein kinase activates the NF-κB transcription factor by dissociating the cytoplasmic NF-κB-IκB complex, *Proc. Natl. Acad. Sci. USA*, **90**:92487–9251.

Libermann, T.A., and Baltimore, D., 1990, Activation of interleukin-6 gene expression through the NF-*kappa*, B transcription factor. *Mol. Cell. Biol.*, **10**:2327–2334.

Lin, Y.-C., Brown, K., and Siebenlist, U., 1995, Activation of NF-κB requires proteolysis of the inhibitor IκB-α: Signal-induced phosphorylation of IκB-α alone does not release active NF-κB, *Proc. Natl. Acad. Sci. USA*, **92**:552–556.

Link, E., Kerr, L.D., Schreck, R., Zabel, U., Verma, I., and Baeuerle, P.A., 1992, Purified IκB-β is inactivated upon dephosphorylation, *J. Biol. Chem.*, **267**:239–246.

Liou, H-C., Nolan, G.P., Ghosh, S., Fujita, T., and Baltimore, D., 1992. The NF-κB precursor, p105, contains an internal IκB like inhibitor that preferentially inhibits p50, *EMBO J.*, **11**:3003–3009.

Lui, M.K., Herrera-Velit, P., Brownsey, R.W., and Reiner, N.E., 1994, CD14-dependent activation of protein kinase C and mitogen-activted protein kinases (p42 and p44) in human monocytes treated with bacterial lipopolysaccharide, *J. Immunol.*, **153**:2642–2652.

Mellitis, K.H., Hay, R.T., and Goodbourn, S., 1993, Proteolytic degradation of MAD(IκBα) and enhanced processing of the NF-κB precursor p105 are obligatory steps in the activation of NF-κB, *Nucleic Acids Res.*, **21**:5059–5066.

Mercurio, F., DiDonato, J., Rosette, C., and Karin, M., 1992, Molecular cloning and characterization of a novel Rel/NF-κB family member displaying structural and functional homology to NF-κB p50/p105, *DNA and Cell Biology*, **11**:523–537.

Mercurio, F., DiDonato, J.A., Rosette, C., and Karin, M., 1993, p105 and p98 precursor proteins play an active role in the NF-κB mediated signal intransduction, *Genes and Development*, **7**:705–718.

Miyamoto, S., and Verma, M., 1995, REL/NF-κB/IκB Story, *Advances in Cancer Research*, **66**:255–292.

Morrison, D.C., and Ryan, J.L., 1987, Endotoxins and disease mechanisms, *Annu. Rev. Med.*, **38**:417–425.

Müller, J.M., Zeigler-Heitbrock, and Baeuerle, P.A., 1993, Nuclear factor kappa B, a mediator of lipopolysaccharide effects, *Immunobiol*, **187**: 233–256.

Neumann, M., Tsapos, K., Scheppler, J.A., Ross, J., Franza, R., Jr., 1992, Identification of complex formation between two intracellular tyrosine kinase substrates: Human c-Rel and p105 precursor of p50 NF-κB, *Oncogene*, **7**:2095–2104.

Nolan, P.G., Ghosh, S., Liou, H.-C., Tempst, P., and Baltimore, D., 1991, DNA binding and IκB inhibition of the cloned p65 subumit of NF-κB, a rel-related polypeptide, *Cell*, **64**:961–969.

Novogrodsky, A., Vanichin, A., Patya, M., Gazit, A., Osherov, N., and Levitzki, A., 1994, Prevention of lipopolysaccharide-induced lethal toxicity by tyrosine kinase inhibitors, *Science*, **264**:1319–1322.

Palombella, V.J., Rando, O.J., Goldberg, A.L., and Maniatis, T., 1994, The ubiquitin-proteosome pathwqay for processing of the NF-κB precursor protein and the activation of NF-κB, *Cell*, **78**:773–785.

Perkins, N.D., Schmid, R.M., Duckett, C.S., Leung, K., Rice, N.R., and Nabel, G.J., 1992, Distinct combinations of NF-κB subunits determine specificity of transcriptional activation, *Proc. Natl. Acad. Sci. USA*, **89**:1529–1533.

Ruben, S., Klement, T., Coleman, T., Maher, M., Chen, C.-H., and Rosen, C., 1992, Isolation of *rel*-related human cDNA that potentially encodes the 65 kD subunit of NF-κB, *Science*, **251**:1490–1493.

Štefanová, I., Hořeješí, V., Ansotegui, I.J., Knapp, W., and Stockinger, H., 1991, GPI-anchored cell surface molecules complexed to protein-tyrosine kinases, *Science*, **254**:1016–1019.

Štefanová, I., Corcoran, M.L., Horak, E.M., Wahl, L.M., Bolen, J.B., and Horak, I.D., 1993, Lipopolysaccharide induces activation of CD14-associated protein tyrosine kinase p53/56[lyn], *Science*, **268**(2):20725–20728.

Taffet, S.M., Singhel, J.F., and Shurtleff, S.A., 1989, Regulation of tumor necrosis factor expression in a macrophage-like cell line by lipopolysaccharide and cyclic AMP, *Cell. Immunol.*, **120**:291–300.

Thompson, J.E., Phillips, R.J., Erdjument-Bromage, Tempst, P., and Ghosh, S., 1995, IκB-β regulates the persistent response in a biphasic activation of NF-κB, *Cell*, **80**:573–582.

Towbin, H., Staehlin, T., and Gorden, J., 1979, Electrophoretic transfer of protein from poly-acrylamide gels to introcellulose sheets: Procedure and some applications, *Proc. Natl. Acad. Sci. USA*, **76**:4350–4354.

Traenckner, E.-M., Wilk, S., and Baeuerle, P.A., 1994, A proteosome inhibitor prevents activation of NF-κB and stabilization of newly phosphorylated form of IκB-α that is still bound to NF-κB, *EMBO J.*, **13**:5433–5411.

Ulevitch, R.J., 1993, Recognition of bacterial endotoxins by receptor-dependent mechanisms, *Advances in Immunology*, **53**:267–289.

Ulevitch, R.J., and Tobias, P.S., 1994, Recognition of endotoxin by cells leading to transmembrane signaling, *Current Opinion in Immunology*, **6**:125–130.

Vincenti, M.P., Burrell, T.A., and Taffet, S.W., 1992, Regulation of NF-κB activity in murine macrophages: effect of bacterial lipopolysaccharide and phorbol ester. *J. Cell. Physiol.*, **150**:204–213.

Watson, R-G., Redmond, H.P., and Boushier-Haynes, D., 1994, Role of endotoxin in mononuclear phaagocyte-mediated inflammatory responses, *J. Leukoc. Biol.*, **56**:95–103.

Weinstein S.L., Sanghera, J.S., Lemke, K., Defranco, A.L., and Pelech, S.L., 1992, Bacterial lipopolysaccharide induces tyrosine phosphorylation and activation of mitogen-activated protein kinases in macrophages, *J. Biol. Chem.*, **267**:14955–14962.

Zheng, S., Brown, M.C., and Taffet, S.M., 1993, Lipopolysaccharide stimulates both nuclear localization of the NF-κB 50 kDa subunit and loss of the 105kDa precursor in RAW264.7 macrophage-like cells, *J. Biol. Chem.*, **268**:17233–17239.

5

Regulation of Inflammatory Cytokine Biosynthesis: Discovery of a Low Molecular Weight Inhibitor and its Molecular Target

JOHN C. LEE*, SANJAY KUMAR and PETER R. YOUNG

1. Introduction

Pro-inflammatory cytokines such as IL-1 and TNF are highly inducible proteins produced by a variety of cell types in response to a wide range of stimuli (Dinarello, 1991). The normally low or non-existent basal mRNA or protein expression level changes rapidly, with an increase in transcription rate and a concurrent increase in mRNA stability and translation efficiency (Beutler, 1986). There is a concomitant increase in intracellular and secreted cytokine protein. The cytokines can be differentially regulated, as evidenced by qualitative differences in response to stimuli as well as sensitivity to various inhibitors.

Cells of the macrophage or monocytic lineage in particular readily respond to bacterial products such as lipopolysaccharide (LPS) to produce a large array of proteins, many of which are secreted and are pharmacologically active, including cytokines, IL-1 and TNF, and a variety of growth factors. They in turn regulate the expression of many other pro-inflammatory molecules such as stromelysin, PLA2, and adhesion molecules. Overproduction of these molecules is often associated with a variety of inflammatory conditions such as rheumatoid arthritis, inflammatory bowel disease, psoriasis, etc. Interestingly, these cytokines can also act in an autocrine fashion, since many of the cytokine receptors are expressed in

*Correspondence: Dr. John C. Lee, Department of Cellular Biochemistry, SmithKline Beecham Pharmaceuticals, P.O. Box 1539, King of Prussia, PA 19406, USA.

these same cells. However, there has been little definitive data on the post "receptor" molecular events. Despite much information about their respective receptors and the way that the cytokines interact with their receptors, very little is known about their signal transduction mechanism.

The first definitive report suggesting that LPS signaling involves protein phosphorylation came from the studies of Han *et al.*, which identified a novel member of the MAP kinase family, p38 kinase (Han *et al.*, 1993; 1994). It has also been observed that IL-1 and TNF are particularly potent activators of this protein kinase (Saklatvala *et al.*, 1993; Freshney *et al.*, 1994; Raingeaud *et al.*, 1995). In characterizing the activation of MAPKAP kinase-2, a substrate of MAP kinase, upon stimulation of cells by insulin or nerve growth factor, Rouse *et al.*, found that it could also be activated by other stimuli such as chemical stress and heat shock (Rouse *et al.*, 1994). Another report also suggests a role for both tyrosine and serine threonine phosphorylation in regulating cytokine production in LPS stimulated macrophages (Weinstein, Jume, and De Franco, 1993). Another member of the same protein kinase family, c-Jun N-terminus protein kinase (JNK), has also been implicated in stress signaling in mammalian cells (Sanchez, 1994; Derijard, 1994; Kyriakis, 1994; Sluss, 1994).

2. Cytokine Suppressive Anti-inflammatory Compounds

Our initial approach was to survey reports in the literature which examined the cytokine suppressive effects of a series of small molecules. The earliest compounds that showed cytokine synthesis inhibition activity were the glucocorticoids (Lee, 1988). An increasing number of low molecular weight compounds have recently been described, although there has been a paucity of information on their mechanisms of action (Lee, 1993). We were encouraged by a report suggesting that lipoxygenase inhibitors, at concentrations higher than those observed for their enzyme inhibitory activities, also inhibited IL-1 production in human monocytes (Dinarello, 1984). A series of novel, imidazole-containing anti-inflammatory compounds, which had been shown to inhibit eicosanoid metabolism in enzyme assays (Lee *et al.*, 1988), also inhibited IL-1 production, but there was no correlation of cytokine suppression activity with inhibition of arachidonate metabolism.

We have been interested in defining the molecular basis of regulation of cytokine biosynthesis in monocytes. A series of low molecular weight compounds were found to inhibit LPS stimulated IL-1 and TNF production in human monocytes (Lee *et al.*, 1988; Lee *et al.*, 1989; Lee *et al.*, 1990; Lee *et al.*, 1993). This compound, when tested at its IC_{50} for cytokine inhibition, had no appreciable effect on DNA, RNA, or protein synthesis. Furthermore, its inhibitory activity on IL-1 production was observed with a number of different stimuli and target cells. Optimal inhibition was observed when the cells were pre-treated or treated early (<2h hr) in the induction phase of IL-1 expression. In addition to IL-1, a few other cytokines were also inhibited, including TNFα, IL-6, IL-8 and gm-CSF, but not g-CSF or IL-1 receptor antagonists. It was subsequently found that these compounds act primarily at the protein level, but not on the cytokine mRNA level (Young, 1993). Western blot analysis demonstrated that the intracellular levels of both cytokines were significantly reduced in treated cells and were not paralleled by similar changes in the respective mRNA, an observation recently confirmed by Perregaux *et al.* (1995). A second effect was to inhibit the release of IL-1 into the medium in response to high concentrations of LPS, an effect mimicked by a variety of lysomotrophic agents such as chloroquine. However this effect is most likely to be compound- but not mechanism of action-specific. Because of the unique properties exhibited by these compounds in inhibiting inflammatory cytokine production, they have been termed cytokine suppressive anti-inflammatory drugs (CSAID^{TM}).

3. Identification of the Molecular Target

While mechanistic studies suggested the level at which cytokine synthesis is regulated, a full description of the way that compounds act is only possible upon identification of the molecular target. Therefore, radiolabelled and radiophotoaffinity ligands were designed and synthesized to help identify, isolate, and sequence the molecular target and clone its cDNA. This approach has been successfully used to identify the molecular targets for the immunosuppressants cyclosporin (Handschumacher *et al.*, 1984) and FK506 (Siekierka *et al.*, 1989; Harding *et al.*, 1989), a 5-lipoxygenase translocation inhibitor (Miller *et al.*, 1990), and the anti-allergy drug Cromoglycate (Hemmerich *et al.*, 1992), and has more recently been extended to the targets for rapamycin and digoxigenin (Brown *et al.*, 1994, Sabatini *et al.*, 1994).

Initially, we synthesized the compound ^{3}H-SB202190 (Fig. 1), and measured its uptake into THP.1 cells, a human monocytic cell line, which upon treatment with LPS showed a pattern of cytokine production and sensitivity to drug inhibition identical to that observed in freshly isolated human peripheral blood monocytes. Uptake of the radioligand was time- and temperature-dependent, saturable and could be competed with unlabelled SB202190, but not with an inactive analogue. These results suggest that a compound-specific binding molecule may exist in THP-1 cell lysates.

A binding assay was subsequently configured to quantitate the binding of ^{3}H-SB202190 to cell lysates, with which the labelled compound was incubated with cytosolic extract of THP.1 cells. The mixture was applied and eluted from G-10 Sephadex, and the bound fraction was eluted in the void volume.

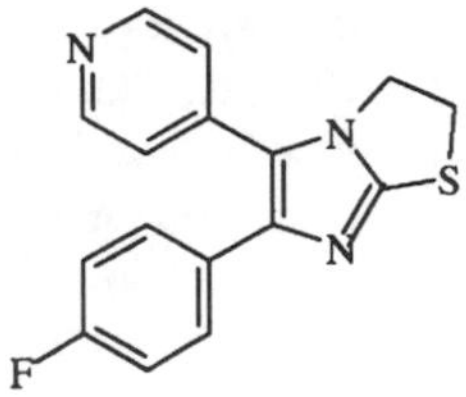

SK&F 86002

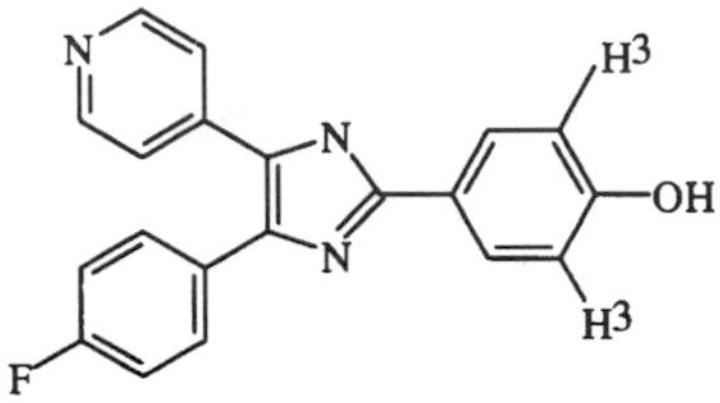

SB 202190

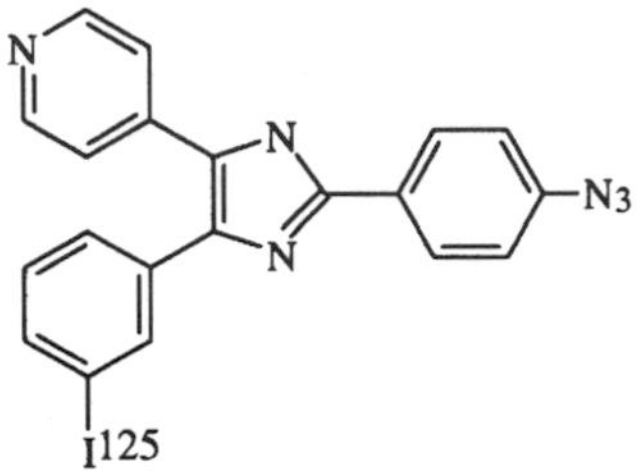

SB 206718

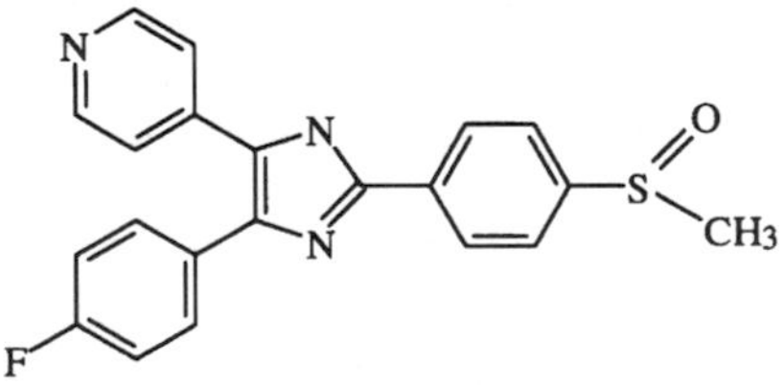

SB 203580

Figure 1 Structures of CSAID compounds.

The binding was specific, time-dependent, reversible, and of high affinity (Kd ≃ 50 nM). More importantly, the rank order potency of many structural analogues exhibited a high degree of correlation between the cytokine biosynthesis inhibition and competition in the binding assay. This assured that the binding activity was relevant to cytokine inhibition. Other structurally or mechanistically unrelated compounds with anti-inflammatory or anti-cytokine activities, such as cyclosporin A, dexamethasone, cyclooxygenase, and lipoxygenase inhibitors, did not compete in the binding assay (Blumenthal *et al.*, unpublished observations), establishing the uniqueness of the target.

Size exclusion chromatography of cell lysates and sensitivity to proteolysis indicated that the molecular target was a protein of ca. 50kDa, which we called CSBP (for CSAID™ binding protein). However, we were unable to purify the target more than about 20-fold as determined by the binding assay, and attempts to affinity purify the protein using compound bound resin were similarly ineffective. Therefore, the radiophotoaffinity analogue ^{125}I-SB206718 was synthesized, which had an IC_{50} of ca. 0.5 uM in the binding assay (Fig. 1). Despite this apparently weak binding affinity, the compound crosslinked to a single 43kDa protein in partially purified cell lysates, which could then be competed with unlabelled active but not inactive compounds in the same rank order of potency as in the binding assay (Lee *et al.*, 1994).

Using a combination of radioligand binding and radiophotoaffinity crosslinking, we were able to purify the protein, subject it to trypsin proteolysis and chemical cleavage by cyanogen bromide, and obtain two unique peptide sequences which were used to make reverse translated synthetic oligonucleotides. These were used to screen a GM-CSF stimulated human monocyte library, from which we obtained a partial cDNA encoding the two peptide sequences. Further screening yielded two cDNAs containing complete open reading frames.

These two cDNAs both contained an open reading frame encoding a protein of 360 amino acids. They were identical in sequence except for an internal region of 75 bp which shared only 43% identity at the nucleotide and protein level (Fig. 2). This suggested that the two proteins were products of alternate splicing; this has recently been confirmed by the sequencing of genomic DNA encoding CSBP. The two protein products were named CSBP1 and 2. Expression of these two proteins in *E. coli* and yeast showed that they could both bind to the radioactive and radiophotoaffinity compounds (Lee *et al.*, 1994).

A search of the GENBANK database revealed that the protein was a novel member of the MAP kinase family of serine-threonine protein kinases. The phylogenetic relationships between the different mammalian MAP kinases are shown in Fig. 3. p38 and mpk2 are the murine and Xenopus homologues of CSBP2, and were discovered at the same time as the CSBPs (Han *et al.*, 1994; Rouse *et al.*, 1994). Subsequently, we have been able to isolated cDNAs for CSBP1 from a murine lymphocyte cDNA library, suggesting that these spliced forms are conserved among mammals (Fig. 3). Indeed, the murine and human forms of CSBP differ by only two amino acids. Recently, another spliced form of CSBP was discovered via its ability to interact with Max, a protein that forms a heterodimer with the transcription factor c-Myc (Zervos *et al.*, 1995). This forms replaces the carboxyterminal 80 amino acids of CSBP2 with an alternate 17 amino acids. The alternate splice forms could affect activation, substrate selectivity, or down regulation by phosphatases.

Chromosomal localization of CSBP indicates that it is on human chromosome 6p21.2, close to or within a region containing the major histocompatibility locus, adjacent to the TNF and heat shock genes (McDonnell *et al.*, 1995). This locus

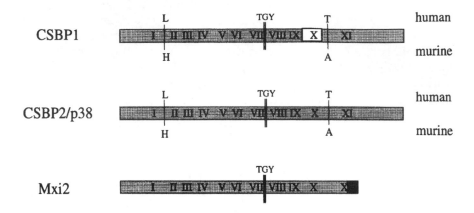

Figure 2. Splice variants of CSBP. The shaded regions represent alternative exons to those in CSBP1. The Roman numerals denote the 11 conserved protein kinase regions identified by Hanks *et al.*, (1989). The amino acids above and below CSBP1 and 2 represent the two amino acid differences found between human and mouse.

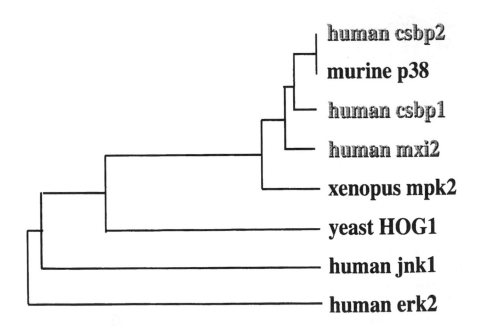

Figure 3 Phylogeny of CSBPs and MAP kinases.

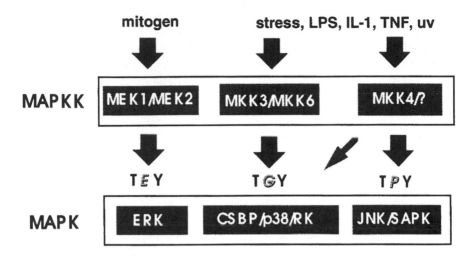

Figure 4. Signaling in the mammalian MAP kinases. This shows the different stimuli leading to activation of the three different MAP kinase members.

has been associated with many inflammatory and autoimmune conditions such as rheumatoid arthritis. Further mapping with determine whether any of these loci are linked to CSBP.

All of the MAP kinases are characterized by an activation loop containing the sequence Thr-Xaa-Tyr, which is phosphorylated on both Thr and Tyr by a MAP kinase in response to extracellular stimuli (Cobb and Goldsmith, 1995). This phosphorylation results in activation of the MAP kinase. There are at least three families of MAP kinases, which differ in the sequence and size of the activation loop: the erks (for extracellular regulated kinases) have a TEY motif; the JNK/SAPKs (for c-Jun NH_2 terminal Kinases or Stress Activated Protein Kinases) have the TPY motif, and the CSBP/p38/RK family has a TGY motif. These variations may be a result of differences in their respective upstream kinases and extracellular stimuli which activate them (Fig. 4) The erks are stimulated by mitogens and several serpentine receptor agonists (Cobb and Goldsmith, 1995), resulting in the activation of ras, raf, MEK (MAP or erk kinase), and erk, which in turn phosphorylates transcription factors such as Elk1 and STAT1, thereby stimulating their transcriptional activity (Gille *et al.*, 1992; Wen *et al.*, 1995). Both the JNKs and CSBPs are activated them (Fig. 4). The erks are stimulated by mitogens and several serpentine receptor agonists (Cobb and Goldsmith, 1995), resulting in the activation of ras, raf, MEK (MAP or erk kinase), and erk, which in turn phosphorylates transcription factors such as Elk1 and STAT1, thereby stimulating their transcriptional activity (Gille *et al.*, 1992; Wen *et al.*, 1995).

Both the JNKs and CSBPs are activated by various stress stimuli including UV, heat, chemical or osmotic shock. IL-1 and TNF, and endotoxin (Derijard *et al.*, 1994; Kyriakis *et al.*, 1994; Han *et al.*, 1994; Rouse *et al.*, 1994; Freshney *et al.*, 1994; Lee *et al.*, 1994; Kracht *et al.*, 1994; Raingeaud *et al.*, 1995). However, more recent studies indicate some differences in the MAP kinase kinases which activate these two MAP kinases, since CSBP is activated by MKK3, MKK4 and MKK6, whereas of these three, only MKK4 activates JNK (Sanchez *et al.*, 1994; Derijard *et al.*, 1995; Lin *et al.*, 1995; Han *et al.*, 1996; Raingeaud *et al.*, 1996).

4. Mechanism of Action of CSAID™ Compounds

Our first priority was to determine how the CSAID™ compounds worked. We were able to show that the compounds inhibited the kinase activity of CSBP whether expressed in *E. coli*, yeast, or mammalian cells (Lee *et al.*, 1994; Cuenda *et al.*, 1995; Kumar *et al.*, unpublished studies). In contrast, there was no inhibition of JNK or erk, MEK, or several other serine-threonine or tyrosine kinases (Cuenda *et al.*, 1995). The ability to inhibit kinase activity followed the same rank order potency as in the binding and cytokine inhibition assays.

The availability of a specific inhibitor of CSBP presented an opportunity to define the downstream events signalled through this enzyme. An immediate candidate was MAPKAP kinase-2, which had previously been found to be an *in vitro* substrate of RK, the rodent homologue of CSBP. Further experiments showed that one of the CSAID™ compounds, SB 203580, was able to inhibit activation of MAPKAP kinase-2 both *in vitro* and *in vivo*, confirming that it was also an *in vivo* substrate (Cuenda *et al.*, 1995). The phosphorylation of hsp27, a small heat shock protein phosphorylated *in vitro* by MAPKAP kinase-2 and known to be phosphorylated *in vivo* by IL-1 treatment, was also inhibited by SB203580, indicating that its phosphorylation was also dependent on the activation of CSBP.

The discovery of an *in vivo* substrate of CSBP allowed us to determine whether SB203580 could also prevent activation of CSBP by binding to the enzyme and preventing its phosphorylation by activating MAP kinase kinases. Using an epitope tagged version of CSBP, we have been able to show that while SB203580 inhibited MAPKAP kinase activation, it did not prevent phosphorylation or activation of CSBP kinase activity (Kumar *et al.*, unpublished data).

We do not presently have an explanation for why there are two different spliced forms of CSBP. Superimposing the CSBP sequence on the recently determined structure of erk2 (Zhang *et al.*, 1994) suggests that this region is not in the substrate binding region indicated by the structure of PKA with its substrate peptide (Zheng *et al.*, 1993). This is consistent with our inability to find any clear differences in substrate specificity between the two forms. However, some experiments in yeast hinted at another possible explanation for the two spliced forms, as described below.

5. Expression of Human CSBP in Yeast

The yeast homologue of CSBP, HOG1, was in fact the first member of the stress activated protein kinases to be discovered since yeast requires it in order to grow under conditions of high osmolarity (Brewster *et al.*, 1993). The pathway by which it is activated *in vivo* is similar to mammalian cells, in that HOG1 requires activation by the MAP kinase kinase PBS2. Because both HOG1 and CSBP can be activated by stress signals, it was of interest to see if CSBP could complement the HOG1Δ phenotype.

Initial attempts to express CSBP2 with a constitutive promoter in a yeast strain deficient of HOG1 failed to yield any transformants (Kumar *et al.*, 1995). However, transformants were obtained if the HOG1 upstream kinase PBS2 was deleted, suggesting that the lack of growth was due to unregulated activation through the HOG1 pathway. This was surprising, since other groups were able to show complementation with p38, the murine homologue of CSBP (Han *et al.*, 1994), and JNK1 (Galcheva-Gargova *et al.*, 1994). Therefore, we expressed both CSBP1 and CSBP2 under the control of an inducible copper metallothionein promoter so that we could independently regulate their expression. We found that CSBP1, but not CSBP2, was able to partially complement the HOG1 deletion. Interestingly, A34V, a mutation in CSBP2, which was introduced serendipitously during cloning, now allowed CSBP2 to complement.

Further analysis of these CSBPs expressed in yeast showed that all three kinases were activated and phosphorylated on tyrosine, but that only CSBP1 and CSBP2 (A34V) activity and phosphorylation were inducible in response to increased osmolarity. This suggested that the inability of CSBP2 to complement was a result of desensitization due to a high basal level of kinase activity, which presumably caused feedback inhibition. Mutations which reduced the kinase activity of CSBP2, such as A34V, restored complementation, and mutations which eliminated the kinase activity, such as K53R or D168A, restored inducibility to the tyrosine phosphorylation but did not complement. Therefore, inducible kinase activity is required for complementation.

We surmise that the different basal levels of kinase activity of CSBP1 and 2 in yeast is the result of differential sensitivity to phosphatases within these cells, since these differences are observed in the presence of equal amounts of protein. We do not know whether the CSBPs show differential sensitivity to phosphatases in mammalian cells, or whether this is a cross-species artifact. The complementation data suggest that the stress activated pathway is conserved across all eukaryotes, although the fine details appear to have changed through evolution.

6. The Role of CSBP in Cytokine Production and Action

The availability of a specific CSBP kinase inhibitor has also allowed us to begin to understand its role in cytokine production and action. In the case of LPS stimulated

IL-1 and TNF, ELISAs and Western blots indicated a 10 to 20-fold reduction in secreted protein levels in the presence of compound. The reduction of mRNA was only 2-fold for TNF, and very little for IL-1, suggesting post-transcriptional regulation. However, much of the inhibition appeared to occur prior to secretion, since there were substantial reductions in the cell associated precursor levels of IL-1 and TNF (Young et al., 1993). Pulse and pulse-chase studies have shown significant inhibition at the protein synthesis step, but not in protein turnover. Similarly, polysome analysis by sucrose gradients seems to indicate a substantial shift of TNF mRNA from polysomes to monosomes upon treatment with inhibitor, suggesting that translation is blocked prior to elongation (Prichett et al., 1995).

These data are consistent with time of action experiments, in which the ability of CSBP inhibitor to block cytokine production from monocytes is measured when added at various times before and after addition of LPS, and is compared to the timing of inhibition by the transcription inhibitor, actinomycin D, and the protein synthesis inhibitor, anisomycin. These data clearly show that the inhibition by CSBP inhibitor is delayed in similar fashion to anisomycin, again supporting the argument that the compounds act at the translational level (Prichett et al., 1995).

It has been established for some time that both IL-1 and TNF can be regulated independently at the translational level (Beutler et al., 1986; Schindler et al., 1990; Kaspar and Gehrke, 1994; Han et al., 1991). In particular, several studies using transfected TNF promoter-5'-UTR-CAT-TNF 3'UTR reporter constructs have established that TNF expression is suppressed at the translational level by AU rich sequences in the 3'URT (Han et al., 1990; Kruys et al., 1989; 1990). Using RAW264.7 cells permanently expressing these constructs, it has been possible to show that inhibition by the CSAID compounds is also dependent on the 3'UTR, so that it can be expected to affect translation of TNF mRNA via a similar mechanism.

While the consequence of inhibiting CSBP in monocytes is to block inflammatory cytokine production, the physiological effects of the inhibitors in other cell types are only now being explored. For example, the inhibition of phosphorylation of the small heat shock protein hsp27 is believed to affect the actin filament organization and function (Zhu et al., 1994; Lavoie et al., 1995). Of more interest is the finding that the CSBP inhibitors suppress the secretion of IL-8 from IL-1 treated endothelial cells by 50%, suggesting that CSBP is not only important in the production of inflammatory cytokines in response to LPS, but also in the action of these cytokines on target cells.

We have described a means by which the naive stress responses of a simple eukaryote such as Saccharomyces cerevisiae have been extended to serve the more complex responses of multicellular organisms such as mammals. When a cell responds to a primary stress such as endotoxin, osmotic, chemical, or heat shock, or UV irradiation, it activates the CSBP MAP kinase pathway, which in turn plays a key role in the production of "stress" cytokines such as IL-1 and TNF. These are released, stimulating the "stress" response in other cells bearing the appropriate receptors (Fig. 5). Since many studies have concluded that

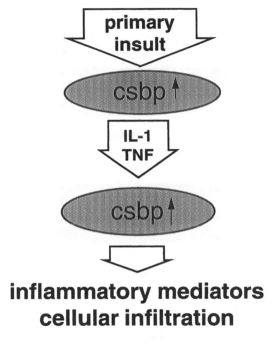

Figure 5. CSBP plays a dual role in mediating signaling events leading to cytokine production and in response to cytokine action.

most stress responses are protective (for example, the heat shock response), this suggests that the inflammatory cytokines can coordinate protective responses both within adjacent cells and by stimulating additional inflammatory mediators. These recruit other cells, such as neutrophils and macrophages, to the inflamed locality, and initiate tissue repair. In normal homeostasis this process is presumably self-limiting, whereas in chronic and acute inflammation these mediators are likely to be overproduced, leading to tissue injury. Hence the evolutionarily conserved intracellular stress responses are linked to extracellular mechanisms for extending the stress response in order to take advantage of local and systemic cellular protective mechanisms.

Acknowledgements

We wish to thank many of our colleagues at SmithKline Beecham and external collaborators for their contribution to this study, B. Metcalf and M. Gowen for their continued interest and support and M. Lark and B. Votta for their review of the manuscript.

References

Beutler, B., Krochin, N., Milsark, I.W., Luedke, C., and Cerami, A, 1986, Control of cachectin (tumor necrosis factor) synthesis: mechanisms of endotoxin resistance. *Science*, **232**:977–980.

Brewster, J.L., V.T. de, N.D. Dwyer, Winter, E., and Gustin, M.C., 1993, An osmosensing signal transduction pathway in yeast. *Science*, **259**:1760–1763.

Brown, E., Albers, M., Shin, T., Ichikawa, K., Keith, C. Lane, W., and Schreiber, S., 1994, *A mammalian protein targeted by G1-arresting rapamycin-receptor complex. Nature*, **369**:756–758.

Cobb, M.H., and Goldsmith, E.J, 1995, How MAP kinases are regulated. *J. Bio. Chem.*, **270**:14843–14846.

Cuenda, A., Rouse, J., Doza, Y.N., Meier, R., Cohen, P., Gallagher, T.F., Young, P.R., and Lee, J.C, 1995, SB 203580 is a specific inhibitor of a MAP kinase homologue which is stimulated by cellular stresses and interleukin-1. *FEBS Lett.*, **364**:229–233.

Derijard, B., Hib, I.M., Wu, I.-H., Barrett, T., Su, B., Deng, T., Karin, M., Davis, and R.J., 1994, JNK1: A protein kinase stimulated by UV light and H-Ras that binds and phosphorylates the c-jun activation domain. *Cell*, **76**:1025–1037.

Derijard, B., Raingeaud, J., Barrett, T., Wu, I.H., Han, J., Ulevitch, R.J., and Davis, R.J., 1995, Independent Human MAP kinase signal transduction pathways defined by MEK and MKK Isoforms. *Science*, **267**:682–685.

Dinarello, C.A., 1991, Inflammatory cytokines: interleukin-1 and tumor necrosis factor as effector molecules in autoimmune diseases. *Curr Opin. Immunol.*, **3**:941–948.

Dinarello, C.A., Bishai, I., Rosenwasser, L., and Coceani, F., 1984, The influence of lipoxygenase inhibitors on the in vitro production of human leukocyte pyrogen and lymphocyte activity factor (IL-1, *Int. J. Immunopharm.*, **6**:43–50.

Freshney, N.W., Rawlinson, L., Guesdon, F., Jones, E., Cowley, S., Hsuan, J., and Saklatvala, J., 1994, Interleukin-1 activates a novel protein kinase cascade that results in the phosphorylation of Hsp27. *Cell*, **78**:1039–1049.

Galcheva-Gargova, Z., Derijard, B., Wu, I-H., and Davis, R.J., 1994, An osmosensing signal transduction pathway in mammalian cells. *Science*, **265**:806–808.

Gille, H., Sharrocks, A.D., and Shaw, P., 1992, Phosphorylation of transcription factor p62TCF by MAP kinase stimulates ternary complex formation at c-fos promoter. *Nature*, **358**:414–417.

Han, J., Lee, J-D., Jiang, Y., Li, Z., Feng, L., and Ulevitch, R.J., 1996, Characterization of the structure and function of a novel MAP kinase kinase (MKK6). *J. Biol. Chem.*, **271**:2886–2891.

Han, J., and Beulter, B., 1990, The essential role of the UA-rich sequence in endotoxin-induced cachectin/TNF synthesis. *Eur. Cytokine Net.*, **1**:71–75.

Han, J., Brown, T., and Beutler B, 1990, Endotoxin-responsive sequences control cachectin/tumor necrosis factor biosynthesis at the translational level. *J. Exp. Med.*, **71**:465–475.

Han, J., Huez, G., and Beutler, B, 1991, Interactive effects of the tumor necrosis factor promoter and 3′-untranslated regions. *J. Immunol.*, **146**:1843–1848.

Han, J., Lee, J.D., Tobias, P.S., and Ulevitch, R.J, 1993, Endotoxin induces rapid protein tyrosine phosphorylation in 70Z B cells expressing CD14. *J. Biol. Chem.*, **268**:25009–25014.

Han, J., Lee, J.-D., Bibb, S.L., Ulevitch, R.J, 1994, A MAP kinase targeted by endotoxin and in mammalian cells. *Science*, **265**:808–811.

Handschumacher, R.E., Harding, M.W., Rice, J., Drugge, R.J., and Speicher, D.W., 1984, Cyclophilin: a specific cytosolic binding protein for cyclosporin A. *Science*, **226**:544–547.

Hanks, S.K., Quinn, A.M., and Hunter, T., 1988, The protein kinase family: conserved features and deduced phylogeny of the catalytic domains. *Science*, **241**:42–52.

Harding, M.W., Galat, A., Uehling, D.E., and Schreiber, S.L., 1989, A receptor for the immunosuppressant FK506 is a cis-trans peptidyl-prolyl isomerase. *Nature*, **341**, 758–760.

Hemmerich, S., Yarden, Y., and Pecht, I., 1992, A cromoglycate binding protein from rat mast cells of a leukemia line is a nucleoside diphosphate kinase. *Biochemistry*, **31**:4574–4579.

Kaspar, R.L., and Gehrke, L., 1994, Peripheral blood mononuclear cells stimulated with C5a or lipopolysaccharide to synthesize equivalent levels of IL-1 mRNA show unequal IL-1β protein accumulation but similar polyribosome profiles. *J. Immunol.*, **153**:277–286.

Kracht, M., Truong, O., Totty, N.F., Shiroo, M., and Saklatvala, J, 1994, Interleukin 1 alpha activates two forms of p54 alpha mitogen-activated protein kinase in rabbit liver. *J. Exp. Med.*, **180**:2017–2025.

Kruys, V., Beutler B., and Huez, G., 1990, Translation control mediated by UA-rich sequences. *Enzymes*, **44**:193–202.

Kruys, V., Marinx O., Shaw G., Deschamps, J., and Huez, G, 1989, Translational blockade imposed by cytokine-derived UA-rich sequences. *Science*, **24**:852–855.

Kumar, S., McLaughlin, M.M., McDonnell, P.C., Lee, J.C., Livi, G.P., and Young, P.R., 1995, Human mitogen-activated protein kinase CSBP1, but not CSBP2, complements a HOG1 deletion in yeast. *J. Biol. Chem.*, **270**:29043–29046.

Kyriakis, J.M., Banerjee, P., Nikolakak, E., Dai, I.T., Rubi, E.A.E., Ahmad, M.F., Avruch, J., and Woodgett J.R, 1994, The stress activated protein kinase subfamily of c-Jun kinases. *Nature*, **369**:156–160.

Lavoie, J.N., Lambert, H., Hickey, E., Weber, L.A., and Landry, J, 1995, Modulation of cellular thermoresistance and actin filament stability accompanies phosphorylation-induced changes in the oligomeric structure of heat shock protein 27. *Mol. Cell. Biol.*, **15**:505–516.

Lee, J.C., Griswold, D.E., Votta, B., and Hanna, H, 1988, Inhibition of monocyte IL-1 production by the anti-inflammatory compound SK&F 86002. *Int. J. Immunopharm.*, **10**:835–843.

Lee, J.C., Rebar, L., and Laydon, J.T, 1989, Effect of SK&F 86002 on cytokine production by human monocytes. *Agents Actions*, **27**(3–4):277–270.

Lee, J.C., Votta, B., Dalton, B.J., Griswold, D.E., Bender, P.E., and Hanna, N., 1990, Inhibition of human monocyte IL-1 production by SKF 86002. *Int. J. Immunotherapy*, **VI**:1–12.

Lee, J.C., Badger, A.M., Griswold, D.E., Dunnington, D., Truneh, A., Votta, B., White, J.R., Young, P.R., and Bender, P.E., 1993, Bicyclic imidazoles as a novel class of cytokine biosynthesis inhibitors. *Ann. N.Y. Acad. Sci.*, **696**:149–170.

Lee, J.C., Laydon, J.T., McDonnell, P.C. Gallagher, T.F., Kumar, S., Green, D., McNulty, D., Blumenthal, M.J., Heys, J.R., Landvatter, S.W., Strickler, J.E., McLaughlin, M.M., Siemens, I.R., Fisher, S.M., Livi, G.P., White, J.R., Adams, J.L., and Young, P.R. 1994, A protein kinase involved in the regulation of inflammatory cytokine biosynthesis. *Nature*, **372**:739–746.

Lee, S.W., Tsou A.P., Chan H., Thomas J., Petrie K., Eugui E.M., and Allison, A.C., 1988, Glucocorticoids selectively inhibit the transcriptin of interleukin-1 gene and decrease the stability of IL-1β mRNA. *Proc. Natl. Acad. Sci. USA*, **85**:1204–1208.

Lin, A., Minden, A., Martinetto, H., Claret, F.-X., Lange-Carter, C., Mercurio, F., Johnson, G.L., and Karin, M., 1995, Identification of a dual specificity kinase that activates the Jun kinases and p38–Mpk2. *Science*, **268**:286–290.

McDonnell, P.C., DiLella, A., Lee, J.C., and Young, P.R., 1995, Localization of the human stress responsive MAP kinase-like CSAID binding protein (CSBP, gene to chromosome 6p21.2/21.3. *Genomics*, **28**:301–302.

Miller, D.K., Gillard, J.W., Vickers, P.J., Sadowski, S., Lercille, C., Mancini, J.A., *et al.*, 1990, Identification and isolation of a membrane protein necessary for leukotiene production. *Nature*, **343**:278–281.

Perregaux, D.G., Dean, D., Cronan, M., Connelly, P., and Gabel, C.A., 1995, Inhibition of interleukin-1 beta production by SKF86002: evidence of two sites of *in vitro* activity and of a time and system dependence. *Mol. Pharmacol.*, **48**:433–442.

Prichett, W., Hand, A., Sheilds, J., and Dunnington D., 1995, Mechanism of action of bicyclic imidazoles defines a translational regulatory pathway of tumor necrosis factor α. *J. Inflamm.*, **45**:97–105.

Raingeaud, J., Gupta, S., Rogers, J.S., Dickens, M., Han, J., Ulevitch, R.J., and Davis, R.J., 1995, Pro-inflammatory cytokines and environmental stress cause p38 mitogen-activated protein kinase activation by dual phosphorylation on tyrosine and threonine. *J. Biol. Chem.*, **270**:7420–7426.

Raingeaud, J., Whitmarsh, A.J., Barrett, T., Derijard, B., and Davis, R.J., 1996, MKK3- and MKK6-regulated gene expression is mediated by the p38 mitogen-activated protein kinase signal transduction pathway. *Mol. Cell. Biol.*, **16**:1247–1255.

Rouse, J., Cohen, P., Trigon, S., Morange, M., Alonso-Llamazares, A., Zamanillo, D., Hunt, T., Nebreda, A.R., 1994, Identification of a novel protein kinase cascade stimulated by chemical stress and heat shock which activates MAP kinase-activated protein MAPKAP kinase-2 and induces phosphorylation of the small heatshock proteins. *Cell*, **78**:1027–1037.

Sabatini, D., Erdjument-Bromage, H., Lui, M., Tempst, P., and Snyder, S., 1994, RAFT1: a mammalian protein that binds to FKBP12 in a rapamycin-dependent fashion and is homologous to yeast TORs. *Cell*, **78**:35–43.

Saklatvala, J., Rawlinson, L.M., Marshall, C.J., and Kracht, M., 1993, Interleukin-1 and tumour necrosis factor activate the mitogen-activated protein (MAP, kinase kinase in cultured cells. *FEBS Lett.*, **334**:189–192.

Sanchez, I., Hughes, R.T., Mayer, B.J., Yee, K., Woodgett, J.R., Avruch, J., Kyriakis, J.M., and Zon, L.I., 1994, Role of SAPK/ERK kinase-1 in the stress-activated pathway regulating transcription factor c-jun. *Nature*, **372**:794–798.

Schindler, R., Gelfand, A., and Dinarello, C.A., 1990, Recombinant C5a stimulates transcription rather than translation of interleukin-1 (IL-1), and tumor necrosis factor: translational signal provided by lipopolysaccharide or IL-1 itself. *Blood*, **76**:1631–1638.

Siekierka, J.J., Hung, S.H.Y., Poe, M., Lin, C.S., and Sigal, N.H., 1989, A cytosolic binding protein for the immunosuppressant FK506 has peptidyl-prolyl isomerase activity but is distinct from cyclophilin. *Nature*, **341**:755–757.

Sluss, H.K., Barrett, T., Derijard, B., and Davis, R.J., 1994, Signal transduction by tumor necrosis factor mediated by JNK protein kinases. *Mol. Cell. Biol.*, **14**:8376–8384.

Weinstein, S.L., Jume, C.H., and DeFranco, A.L., 1993, Lipopolysaccharide-induced protein tyrosine phosphorylation in human macrophages is mediated by CD14. *J. Immunol.*, **151**:3829–3838.

Wen, Z., Zhong, Z., and Darnell, J.E. Jr., 1995, Maximal activation of transcription by Stat1 and Stat3 requires both tyrosine and serine phosphorylation. *Cell*, **82**:241–250.

Young, P.R., McDonnell, P., Dunnington, D., Hand, A., Laydon, J., and Lee, J.C., 1993, Bicyclic imidazoles inhibit IL-1 and TNF production at the protein level. *Agents Actions*, **39**:C67–C69.

Zervos, A.S., Faccio, L., Gatto, J.P., Kyriakis, J.M., and Brent, R., 1995, Mxi2, a mitogen-activated protein kinase that recognizes and phosphorylates Max protein. *Proc Natl. Acad. Sci. USA*, **92**:10531–10534.

Zhang, F., Strand, A., Robbins, D., Cobb, M.H., and Goldsmith, E.J., 1994, Atomic structure of the MAP kinase ERK2 at 2.3 A resolution. *Nature*, **367**:704–710.

Zheng, J., Knighton, D.R., Ten Eyck, L.F., Karlsson, R., Xuong, N-H., Taylor, S.S., and Sowadski, J.M., 1993, Crystal structure of the catalytic subunit of cAMP-dependent protein kinase complexed with MgATP and peptide inhibitor. *Biochemistry*, **32**:2154–2161.

Zhu, Y., O'Neill, S., Saklatvala, J., Tassi, L., and Mendelsohn, M.E., 1994, Phosphorylated HSP27 associates with the activation-dependent cytoskeleton in human platelets. *Blood*, **84**:3715–3723.

6

Signaling by the Cytokine Receptor Superfamily

JAMES N. IHLE

1. Introduction

Hematopoiesis is regulated through the availability of appropriate cytokines, which control a variety of cellular functions including cell cycle progression, apoptosis, differentiation, and functional cellular responses. On one hand, the cytokines involved exhibit extensive pleiotropy and may have quite different functions on cells of different lineages. Conversely, a variety of cytokines may have virtually identical effects on cells at a particular stage of development or differentiation. Over the past few years, numerous aspects of the receptors involved and the signaling pathways activated have begun to provide the information necessary to explain the complex biology associated with hematopoietic cytokines.

Most hematopoietic cytokines function through interaction with receptors of the cytokine receptor superfamily. This family of receptors is not exclusively utilized by the hematopoietic growth factors, but rather contains members that are important to the function of a variety of cell lineages (Bazan, 1990). The cytokine receptor superfamily is characterized by an extracellular domain of approximately 200 amino acids with structural similarity. The type I cytokine receptor subfamily is characterized by four positionally conserved cysteine residues and a WSXWS motif in the extracellular domain. The type II cytokine receptor subfamily, made up of the interferon (IFN) receptors, contains characteristic cysteine pairs at both the amino and terminal regions of the extracellular domain. The predicted overall structures of the extracellular domains of the type I and II subfamilies are similar and consist of two repeated domains that contain seven β strands folded in an

Telephone: +1 901 522 0422; Fax: +1 901 525 8025

anti-parallel fashion to form a barrel-like structure. The crystallography structure of the growth hormone receptor (De Vos *et al.*, 1992; Somers *et al.*, 1994) indicates that the ligand is bound in the hinge region between the two domains. Based on the predicted structures, it is proposed that the type I and II receptors evolved from a common progenitor with evolutionary links to a 100-amino acid fibronectin type III domain found in cell surface proteins with adhesive functions.

The cytokine receptor superfamily members can be further divided into subfamilies based on the number of receptor chains and the utilization of common signaling chains. The receptors for Epo (Youssoufian *et al.*, 1993), growth hormone (Leung *et al.*, 1987), prolactin (Ebert *et al.*, 1994), G-CSF (Fukunaga *et al.*, 1993), and thrombopoietin (Bartley *et al.*, 1994; Kaushansky *et al.*, 1994) consist of single ligand binding chains containing cytoplasmic domains that are critical for signal transduction. The subfamily consisting of the receptors for IL-6, oncostatin M (OSM), ciliary neurotropic factor (CNTF) leukemia inhibitory factor (LIF), and IL-11 utilize ligand binding chains that associate with gp130 or the highly related LIFRβ signaling chains (Stahl and Yancopoulos, 1993). The IL-12 receptor utilizes a gp130 related chain and requires at least one additional chain that has yet to be cloned (Chua *et al.*, 1994). The IL-2 receptor subfamily consists of the receptors for IL-2 (Minami *et al.*, 1993; Leonard *et al.*, 1994), IL-4 (Leonard *et al.*, 1994), IL-7 (Noguchi *et al.*, 1993), IL-9 (Renauld, Druez *et al.*, 1992), and IL-15 (Giri, Ahdieh *et al.*, 1994). All of these receptors utilize a common γ_c chain and a specific α chain related to the IL-2 receptor β chain, with the exception of IL-15, which utilizes a IL-2R β chain (Giri *et al.*, 1994). In addition, IL-13 may utilize receptor chains shared with the IL-4 receptor which, while originally proposed as the γ_c chain, may be a related receptor chain (Obiri *et al.*, 1995; Zurawski *et al.*, 1993). The IL-3/GM-CSF/IL-5 receptor subfamily utilizes a ligand specific binding α chain that associates with a common β_c chain (Miyajima *et al.*, 1993). The IFNα/β and IFNγ receptors each utilize at least two distinct chains (Soh *et al.*, 1994; Novic *et al.*, 1994). The cytoplasmic domains of both chains are essential for signal transduction. Finally, the IL-10 receptor consists of a ligand binding chain with structural similarity to the IFN receptors (Tan *et al.*, 1993; Ho *et al.*, 1993), and is hypothesized to require a second, unidentified chain similar to the IFN receptors.

2. Cytokine Receptor Oligomerization and Induction of Tyrosine Phosphorylation

The first critical ligand-induced event is the dimerization/oligomerization of the receptor components. In the case of the single chain receptor for growth hormone, this occurs through the ability of the single receptor chain to bind two sites to the ligand (De Vos *et al.*, 1992). In cases such as the Epo receptor this may be mediated through ligand dimers. The importance of dimerization is illustrated by the constitutive activation of the Epo receptor by an Arg → Cys mutation, which results in constitutive receptor aggregation through the formation of disulfide-linked homodimers (Youssoufian *et al.*, 1993). In the case of the IL-6 receptor, the primary function of the ligand binding component is to associate with and cause the aggregation of gp130 (Murakami *et al.*, 1993), as exemplified by the ability of the soluble complex to activate signal transduction. The importance of dimerization/oligomerization of the cytoplasmic domains of the cytokine receptors has been particularly evident in a variety of studies utilizing chimeric receptors. In these studies various extracellular domains have been utilized to allow antibody or ligand dependent aggregation of cytoplasmic domains of cytokine receptors.

Although members of the cytokine receptor superfamily lack kinase domains or other catalytic motifs, a variety of studies indicate that they couple ligand binding to the induction of protein tyrosine phosphorylation. The importance of tyrosine phosphorylation was initially indicated by the observation that mutations of the receptors, which eliminated functional activity, similarly eliminated the ability to couple ligand binding to induction of the tyrosine phosphorylation. Substrates of ligand-induced tyrosine phosphorylation include one or more of the receptor chains, suggesting the close association of a kinase. Based on these observations, it was hypothesized that the cytokine receptor superfamily members associate with cytoplasmic kinases that are activated following ligand binding. In order to further explore this hypothesis and identify potential candidate kinases, a variety of studies have examined the ability of cytokines to affect the state of phosphorylation and/or catalytic activity of cytoplasmic protein tyrosine kinases.

3. The Role of Src Family Kinases in Signaling by the Cytokine Receptor Superfamily

The potential involvement of members of the src family of tyrosine kinases had considerable appeal as their role in signaling through the T cell receptor became apparent (Weiss and Littman, 1994). Hematopoietic cells express several of the src-related kinases, including Src, Lck, Lyn, Fyn and Hck, and a few reports have implicated some of these kinases in cytokine signaling. Specifically, one group has implicated Lyn in IL-3 signaling (Torigoe *et al.*, 1992). We have examined the effects of IL-3 on Lyn activity and tyrosine phosphorylation in a variety of IL-3 dependent cell lines and have been unable to demonstrate any alterations, suggesting that Lyn activation may be limited to specific cell lines or perhaps stages of differentiation. Similarly, one group has implicated Lyn in signaling through the G-CSF receptor (Corey *et al.*, 1994).

In contrast to the paucity of studies dealing with src kinases and most cytokine receptors, there have been a number of studies dealing with the potential role of Lck, Lyn, or Fyn in IL-2 receptor function (Minami *et al.*, 1993; Taniguchi, 1995;

Taniguchi and Minami, 1993). The evidence for Lck involvement is based on the observation that Lck *in vitro* kinase activity is increased two- to five-fold following IL-2 binding and association with the IL-2 receptor β chain. The basis for the activation of catalytic activity is not known, although it has been speculated to be due to either the loss of tyrosine phosphorylation of a site that negatively affects activity and/or the phosphorylation of tyrosine residues within the activation loop of the catalytic domain. Lck, as well as Lyn and Fyn, have been shown to associate with an acidic domain within the cytoplasmic domain of the IL-2 receptor β chain. Curiously, this association does not require the Lck SH2 domain, but rather requires a region in the kinase catalytic domain. Deletions of the acidic domain do not affect the mitogenic activity of the receptor, but they do affect the activation of cHa-ras and the induction of c-fos and c-jun. This suggests a role for Lck in the activation of the ras signaling pathway and, perhaps through its effects on the ras pathway, gene induction. Whether src-related kinases participate in other receptors of the IL-2 receptor subfamily is largely unknown, although one study (Venkitaraman and Cowling, 1992) implicates Fyn in IL-7 receptor function.

4. The Role of Tec, Fes and Zap70 Family Kinases in Signaling by the Cytokine Receptor Superfamily

In addition to the src kinases, recent reports have suggested a role for members of the Tec family kinases. The Tec family consists of five related cytoplasmic kinases: Tec, Btk, Itk, Bmx and Txk. The Tec kinases have structures that are very similar to the src subfamily of protein tyrosine kinases, but they lack sites for myristalation and a carboxyl regulatory site of tyrosine phosphorylation. In the amino terminal region there is homology to the pleckstrin homology domain (PH) and a characteristic proline rich region. Mutations in Btk are associated with hypogammaglobulinanemia, suggesting a role in B cell signaling (Tsukada *et al.*, 1993; Rawlings *et al.*, 1993; Thomas *et al.*, 1993). Recent studies have suggested that Tec is activated following IL-3 stimulation (Mano *et al.*, 1995). However, a second study failed to detect activation in response to IL-3 but did find association of Tec with c-kit and activation of Tec following stimulation of cells with stem cell factor (Tang *et al.*, 1994). In another study, Tec, as well as Btk, was found to associate with gp130 and to be activated following IL-6 stimulation (Matsuda *et al.*, 1995b). It will obviously be important to determine whether the activation of members of this family is seen consistently with these and other cytokines.

Syk and Zap70 constitute a unique family of proteins tyrosine kinases that are characterized by the presence of two SH2 domains. Extensive studies have demonstrated their central role in signaling through the T and B cell receptors (Weiss and Littman, 1994). In addition, Syk has been implicated in the signaling events associated with platelet activation (Rivera and Brugge, 1995). Recruitment to the TCR or BCR complex occurs through the interaction of SH2 domains with tyrosine phosphorylated sites with antigen recognition activation motifs (ARAMs). A recent study found that Syk associates with the IL-2 receptor β chain through the membrane proximal serine rich region, and that association does not require ligand activation (Minami *et al.*, 1995). Since the serine rich region does not contain sites for tyrosine phosphorylation, it is hypothesized that Syk associates with the receptor complex in a unique manner, perhaps not requiring the SH2 domains. It has also been reported that Syk is involved in signaling by the GM-CSF receptor and forms a complex with Lyn (Corey *et al.*, 1994). It is not known whether Syk associates with any of the other cytokine receptor superfamily members.

Recent studies (Hanazono *et al.*, 1993a,b) have suggested that Fes is activated in response to GM-CSF, IL-3, and Epo (Hanazono *et al.*, 1993a,b) and associates with the receptors. In contrast, we have not found that IL-3 and Epo have any effect on the level of Fes tyrosine phosphorylation or on the *in vitro* kinase activity in a variety of growth factor-dependent cell lines.

5. The Role of Janus Kinases in Signaling by the Cytokine Receptor Superfamily

The most consistently implicated family of protein tyrosine kinases in cytokine signaling has been the *Janus* protein tyrosine kinases (Jaks). The Jaks were identified during a period when numerous new protein tyrosine kinases were being cloned by homology approaches (Ziemiecki *et al.*, 1994; Ihle *et al.*, 1994, 1995). Their structure is unique in containing a carboxyl kinase domain and, immediately amino terminal, a psuedokinase domain. The Jaks contain no Src homology (SH) domains nor any other previously identified protein motifs. Among the family members there are regions of homology that have been referred to as Jak homology (JH) domains. The mammalian family currently consists of four members: Jak1, Jak2, Jak3, and tyrosine kinase 2 (Tyk2), which vary in size from 120 kDa to 140 kDa. Jak1, Jak2, and Tyk2 are widely expressed, while Jak3 is primarily expressed in hematopoietic lineages where its levels are affected by T cell activation (Kawamura *et al.*, 1994), macrophage activation (Kawamura *et al.*, 1994), and terminal differentiation of granulocytes (Rane and Reddy, 1994).

The murine Jaks have been genetically mapped by interspecific hybrids (O. Silvennoinen, N., Jenkins, N. Copleland, and J.N. Ihle). The *Jak1* gene is very tightly linked to *Pgm2* on chromosome 4, which would correspond to human chromosome band 1p22.1, consistent with the human mapping data (Pritchard *et al.*, 1992). The *Jak2* gene is genetically linked to *Fas* on chromosome 19, which would correspond to human 10q23-q24.1, although the human gene had been previously mapped to 9p24 (Pritchard *et al.*, 1992). The basis for this discrepancy is not known, but it may have arisen from using the murine cDNA to localize the human gene in the initial studies. The *Jak3* gene is located distal of *JunD* in the middle of chromosome 8, a region that has homology with human chromosome 19p13. Interestingly, this would place *Jak3* very near the human *Tyk2* gene at 19p13.2 (Firmbach-Kraft, Byers *et al.*, 1990).

Although initially identified in mammalian species, a Jak homology has been identified in *Drosophila* (Binari and Perrimon, 1994) as the gene associated with the *hopscotch* (*hop*) mutations. Within the carboxyl-kinase domain, the *Drosophila* gene is 39% identical to Jak1/Jak2 or Tyk2. Within the kinase-like domain the identity with Jak1/Jak2 or Tyk2 is 27%, 24%, or 21%, while the amino terminal region has identities of 19%, 23% and 20% respectively. The maternal *hop* product is required for the proper levels of expression of particular stripes of pair-rule genes. In contrast, the *Drosophila Tumorous-lethal* gene is a dominant mutation of the *hopscotch* locus (Hanratty and Dearolf, 1993). This mutation causes the abnormal proliferation and differentiation of the larval hematopoietic system (fly leukemia), leading to late larval/pupal lethality. The mutation is a single amino acid change, Gly^{341} to Glu, in the amino terminal region, which affects kinase activity through an unknown mechanism (Luo *et al.*, 1995). It can be speculated that the mutation, which is not in a conserved region of the gene, affects the Jak dimerization directly or through altered association with a cellular binding protein.

6. Utilization of Jaks in Cytokine Signaling

The Jaks have been implicated in cytokine signaling by the demonstration of ligand induced activation of *in vitro* kinase activity in immune precipitation kinase reactions and/or induced tyrosine phosphorylation. In addition, as discussed below, the Jaks associate with one or more chains of cytokine receptors. In many studies the activation of Jak catalytic activity is quite dramatic, often in excess of a fifty-fold increase as compared to the published data relating to the other cytoplasmic protein tyrosine kinase, in which activation of kinase activity is generally on the order of two- to five-fold. Using the criteria of activation of kinase activity and/or tyrosine phosphorylation, all cytokines that utilize receptors of the cytokine receptor superfamily have been shown to utilize one or more of the Jaks. A summary of the data is presented in Table I.

The first indication that the Jaks were involved in signaling through the cytokine receptor superfamily came from studies to define the genes involved in IFN signaling with mutant cell lines selected for loss of signaling (McKendry *et al.*, 1991; John *et al.*, 1991; Pellegrini *et al.*, 1989). Using an expression cloning approach, a gene was cloned that had the ability to restore the IFN response of one of the complementation groups (Velazquez *et al.*, 1992). Sequencing of the gene identified it as Tyk2. The significance of this observation in relation to cytokine signaling became evident with the demonstration that Jak2 associated with the

Table I. Jak Activation in Cytokine Responses.

Cytokine	Jak Activation (Receptor Association)	References
IL-2	Jak1 (β chain); Jak3 (γ_c chain)	(Johnston, Kawamura et al., 1994; Witthuhn, Silvennoinen et al., 1994)
IL-3	Jak2 (β_c chain)	(Silvennoinen, Witthuhn et al., 1993)
IL-4	Jak1, Jak3 (γ_c)	(Witthuhn, Silvennoinen et al., 1994)
IL-5	Jak2 (β_c chain)	(Sato, Katagiri et al., 1994)
IL-6	Jak1, Jak2, Tyk2 (gp130)	(Stahl, Boulton et al., 1994; Narazaki, Witthuhn et al., 1994)
IL-7	Jak1, Jak3 (γ_c chain)	(Zeng, Takahashi et al., 1994; Nabavi, Grusby et al., 1990)
IL-9	Jak1, Jak3	(Lew, Decker et al., 1989)
IL-10	Jak2, Tyk2	(Kakkis, Riggs et al., 1989)
IL-11	Jak2 (gp130)	unpublished data
IL-12	Jak2; Tyk2	(Bacon, McVicar et al., 1995)
IL-15	Jak1 (β chain), Jak3 (γ_c chain)	unpublished data
LIF	Jak1, Jak2, Tyk2 (LIFRβ)	(Stahl, Boulton et al., 1994)
OSM	Jak1, Jak2, Tyk2 (gp130)	(Stahl, Boulton et al., 1994)
CNTR	Jak1, Jak2, Tyk2 (gp130)	(Stahl, Boulton et al., 1994)
Epo	Jak2	(Witthuhn, Quelle et al., 1993)
G-CSF	Jak1, Jak2	(Nicholson, Oates et al., 1994; Shimoda, Iwasaki et al., 1994)
GM-CSF	Jak2 (β_c)	(Quelle, Sato et al., 1994)
Tpo	Jak2	(Drachman, Griffin et al., 1995; Wei, Charles et al., 1995)
IFN-α/β	Jak1 (β chain), Tyk2 (α chain)	(Muller, Broscoc et al., 1993; Velazquez, Fellous et al., 1992)
IFN-γ	Jak1 (α chain), Jak2 (β chain)	(Muller, Briscoc et al., 1993; Watling, Guschin et al., 1993)
GH	Jak2	(Artgetsinger, Campbell et al., 1993)
Prolactin	Jak2	(Rui, Kirken et al., 1994; Lebrun, Ali et al., 1994)

Epo receptor and was inducibly tyrosine phosphorylated and activated by Epo (Witthuhn *et al.*, 1993); that Jak2 was activated in the response to growth hormone (Artgetsinger *et al.*, 1993); and that Jak2 was activated in the response to IL3 (Silvennoinen *et al.*, 1993). Subsequently, and very rapidly, the activation of Jaks was found in the response to other cytokines. In particular, using the IFN signaling deficient mutants, it was demonstrated that the response to IFNα is dependent upon the presence of both Jak1 and Tyk2 (Muller *et al.*, 1993), while the response to IFNγ is dependent upon Jak1 and Jak2 (Watling *et al.*, 1993; Muller *et al.*, 1993).

The IL-6 subfamily of cytokines was shown to activate multiple Jaks, depending to some extent on the cell lines examined (Stahl *et al.*, 1994; Narazaki *et al.*, 1994). Prolactin, like growth hormones, was shown to activate Jak2 (Campbell *et al.*, 1994; Rui *et al.*, 1994; Dusanter-Fourt *et al.*, 1994; DaSilva *et al.*, 1994). G-CSF has been shown to activate both Jak1 and Jak2 (Nicholson *et al.*,

1994; Shimoda *et al.*, 1994). IL-12, which utilizes a gp130 related signaling chain, activates Jak2 and Tyk2 (Bacon *et al.*, 1995). An IL-2 subfamily of cytokines, which utilizes an IL-2 receptor γ_c chain, activates Jak1 and Jak3 (Witthuhn *et al.*, 1994; Zeng *et al.*, 1994; Johnston *et al.*, 1994; Kirken *et al.*, 1994; Tanaka *et al.*, 1994; Yin *et al.*, 1994). It should be noted that those cytokines that activate a single Jak invariably activate Jak2, while many receptor systems activate two Jaks. This could be interpreted to indicate that, in cases other than Jak2, a co-dependency may exist for activation; this is discussed below.

7. Association of Jaks with Cytokine Receptor Chains and Activation by Receptor Aggregation

In addition to cytokine induced tyrosine phosphorylation and activation of kinase activity, a variety of studies have demonstrated the physical association of Jaks with cytokine receptors. The single chains of the receptors for Epo (Witthuhn *et al.*, 1993; Miura *et al.*, 1994), growth hormone (VanderKuur *et al.*, 1994) and prolactin (DaSilva *et al.*, 1994; Dusanter-Fourt *et al.*, 1994; Lebrun *et al.*, 1994) associate with Jak2 through a membrane proximal region that contains the box1 and box2 homology domains. Specifically, deletions or specific mutations within this area disrupt Jak2 association. This association has been found to be either constitutive or ligand dependent, depending upon the approaches used to examine association. In the case of ligand induced association, tyrosine phosphorylation of the receptor is not required. As with other receptor associations described below, this can be interpreted to indicate that there is a relatively high affinity of association that exists in the absence of ligand, but that receptor aggregation may increase the affinity or decrease the dissociation rate of Jak2 from the receptor complex. Based on these observations, it is hypothesized that receptor aggregation is critical for bringing the Jaks in sufficient proximity to allow phosphorylation, most likely in *trans*, at the activation site within the activation loop of the kinase catalytic domain. In the case of Jak2, one of the major sites of tyrosine phosphorylation is in the KEYY motif (F.W. Quelle and J.N. Ihle, unpublished data). Consistent with this, activation of the Epo receptor by the Arg $\rightarrow$ Cys mutation is associated with constitutive, ligand-independent activation of Jak2.

The cytokine receptors for IL-3, GM-CSF, and IL-5 utilize a common β_c chain and a unique, ligand binding α chain. Using chimeric receptors, it has been demonstrated that the cytoplasmic domain of the β_c chain alone can support signaling, while the cytoplasmic domain of the α chain cannot (Weiss *et al.*, 1993; Eder *et al.*, 1994). The β_c chain binds Jak2, and the membrane proximal region is necessary and sufficient for activation.

The use of chimeric receptors has shown that the cytoplasmic domains of both the IL-2 receptor β and γ_c chains are both required for signal transduction (Nelson *et al.*, 1994; Watowich *et al.*, 1994). The β chain specifically associates with Jak1 through the membrane proximal region containing the box1/box2 motifs, while Jak3 associates with γ_c chain (Russell *et al.*, 1994; Miyazaki *et al.*, 1994). Mutations or truncations of the γ_c chain that affect signal transduction, which in humans are associated with X-linked combined immunodeficiency, disrupt the ability of Jak3 to associate with the receptor.

The group of cytokines that includes IL-6, oncostatin M (OSM), ciliary neurotropic factor (CNTF), leukemia inhibitory factor (LIF), and IL-11 utilize gp130, or the highly related LIFRβ chain, for signal transduction (Stahl and Yancopoulos, 1993). This group of cytokine is unique in the ability to activate multiple Jaks (Stahl *et al.*, 1994; Narazaki *et al.*, 1994). As with the other receptors, association requires the membrane proximal region of the cytoplasmic domain, which contains the box1/box2 motifs. Although they activate several Jaks, there is no co-dependency for their activation (Guschin, Rogers *et al.*, 1995) like that seen with the IFN receptors (Muller *et al.*, 1993; Watling *et al.*, 1993). However, the absence of Jak1 uniquely eliminates downstream signaling events, including gp130 and Stat tyrosine phosphorylation. Moreover, a dominant negative Jak2 can suppress the activation of Jak1 and downstream signaling. It is hypothesized from the data that either Jak2 or Tyk2 is required in the receptor complex for the efficient activation of Jak1, and that Jak1, because of substrate specificity or association, is uniquely able to phosphorylate gp130, which is required for the subsequent recruitment and phosphorylation of Stats (discussed below).

The ability of various chains of the IFN receptors to associate with Jaks has been examined, although none of the studies have defined the domains with which the Jaks associate within the receptors. In the case of the IFNα/β receptor, Tyk2 associates with the first receptor chain cloned (α chain) (Colamonici *et al.*, 1994), while Jak1 associates with the recently cloned β chain (Novick *et al.*, 1994). In the IFNγ receptor, Jak1 associates with the α chain (Igarashi *et al.*, 1994), while Jak2 associates with the recently cloned β chain (in preparation). Importantly, the absence of any one of the kinases results in the lack of activation of the other kinase (Muller *et al.*, 1993; Watling *et al.*, 1993), indicating an absolute co-dependency for two Jaks in each case. It is hypothesized that this co-dependency is due to ligand-induced heterodimerization of the receptors and Jaks.

In summary, the above data suggest that there are numerous variations on a common theme involving ligand/receptor mediated Jak activation. The simplest model is one in which a single receptor chain associates with a single Jak and ligand binding induces Jak association homodimerically. Interestingly, the only Jak involved in this type of response is Jak2. A second variation involves the ability of a single receptor chain to associate with multiple Jaks, as in the case of gp130 or G-CSE, consequently introducing the possibility of heterodimeric complexes. A third variation involves the binding of distinct Jaks by two receptor chains such as occurs in the case of IL-2 family of receptors and the IFNs. In this case there is an obligate Jak heterodimerization that is essential for activation of either Jak.

The obligate requirement for multiple Jaks suggests the possibility that individual Jaks may differ in the efficiency of trans- or autophosphorylation *in vivo* in the context of the receptor complex. It should be noted, however, that over-expression of any of the Jaks, individually, in COS cells or in insect cells results in their activation (F.W. Quelle, B. Witthuhn and J.N. Ihle, unpublished observations). Nevertheless, Jak1 may be very inefficient *in vivo* in autophosphorylation, and may require another Jak for phosphorylation in the activation loop, as described above in the case of IL-6 signaling. This type of requirement has been shown for the TGFβ type I and II serine/threonine receptors (Attisano *et al.*, 1994).

The domains in the Jaks that are responsible for receptor association have been more difficult to define. Using complementation assays in Jak2 negative cell lines, deletions that remove conserved domains of the amino terminal region or the pseudokinase domain of Jak2 eliminate the ability to functionally reconstitute IFNγ responses (Witthuhn *et al.*, unpublished data). Similarly, deletion of the pseudokinase domain of Tyk2 eliminates its ability to complement Tyk2 deficient cells for Jak1 or Stat tyrosine phosphorylation or activation (Velazquez *et al.*, 1995).

Ligand binding induces tyrosine phosphorylation and activation of *in vitro* catalytic activity. In the case of the Epo induced phosphorylation of Jak2, phosphorylation occurs at many sites that have yet to be identified. In these cases, it is hypothesized that ligand binding and the subsequent receptor aggregation that occurs brings the Jaks in sufficient proximity and density to allow trans autophos-phorylations to occur. In the response to Epo, Jak2 tyrosine phosphorylation occurs on multiple sites but includes phosphorylation of the DKYY sequence that occurs in the proposed activation loop of the kinase domain.

8. The Requirements for Jaks in Cytokine Signaling

Several observations support the hypothesis that Jaks play a key role in cytokine signaling. Perhaps the strongest evidence has come from studies with the IFN signaling mutants. These studies demonstrate that the absence of Jak1 can eliminate all signaling by both the IFNα/β and the IFNγ receptor complexes (Muller *et al.*, 1993). The absence of Jak2 or Tyk2 eliminates signaling by the IFNα/β receptors complex or the IFNγ receptor complex respectively (Watling *et al.*, 1993; Velazquez *et al.*, 1992). In addition, mutants of cytokine receptors, which uncouple Jak activation from ligand binding, invariably eliminate all detectable signaling, including the activation of other receptor-associated protein tyrosine kinases. Finally, recent studies have demonstrated that dominant negatives of Jak2 can suppress the mitogenic response of cells to Epo (Zhuang *et al.*, 1994).

The role of Jak3 in lymphoid development has been supported further by gene disruption and the existence of mutations of Jak3 in humans. Using homologous recombination in embryonic stem cells to disrupt the genomic locus, mice have been obtained that are unable to produce a functional Jak3. One study targeted the disruption to the 5' end of the gene (Vogel and Fujita, 1995), while a second study targeted the kinase catalytic domain (Krosl *et al.*, 1995). Residual protein was not detected in either case. In both studies, Jak3-deficient mice were found to have profound reductions in thymocytes and severe B-cell and T-cell lymphopenia, similar to that seen in severe combined immunodeficiencies (SCID). With age, some phenotypically mature T-cells and B-cells were found, although these cells were functionally deficient and failed to respond to a variety of mitogenic stimuli. Importantly, the deficiencies included responses that would not normally require Jak3, demonstrating that the absence of Jak3 affected the differentiation of the cells.

The phenotype of Jak3-deficient mice is comparable to that seen in mice in which there is disruption of either the receptor γ_c chain (Rolling *et al.*, 1995), the IL-7 receptor α chain (Brunn *et al.*, 1995) or the IL-7 gene (Matsuda *et al.*, 1995a). Since IL-7 activates Jak3 (Zeng *et al.*, 1994), presumably through the use of the γ_c receptor component, it can be hypothesized that IL-7 regulates a critical, non-redundant stage in early lymphoid development through the activation of Jak3. It should be noted that IL-7 has also been shown to activate members of the src family of kinases in pre-B cells (Seckinger and Fougereau, 1994).

In humans, one cause of SCID is mutations of the γ_c chain of the IL-2 receptor on the X chromosome (Leonard *et al.*, 1994). However, X-linked SCIDs account for only about half of the human SCIDs, indicating the involvement of other loci. Recently, three SCID patients have been shown to have mutations in Jak3; they fail to express Jak3 (Gollob and Coffman, 1994; Dong *et al.*, 1995). The phenotype of these cases is generally consistent with that seen in Jak3-deficient mice. The human cases have phenotypically mature B-cells, although, as in mice, the B-cells are functionally defective. It is important to determine the frequency of human SCID due to Jak3 mutations and to develop gene therapy approaches that might be used for these patients.

9. Receptor Phosphorylation and Docking of Signaling Proteins

The response to most cytokines is associated with the tyrosine phosphorylation of one or more of the cytokine receptor chains, although in very few cases have the *in vivo* sites of phosphorylation been defined. It can be hypothesized that the activated Jaks are responsible for receptor tyrosine phosphorylation on some, if not all, sites. Consistent with this hypothesis, the co-expression of Jaks with receptors in insect cells results in tyrosine phosphorylation, although it has not been shown that the same sites are phosphorylated. Tyrosine phosphorylation of gp130 is induced in response to IL-6. Phosphorylation is dramatically reduced in cells that lack Jak1, further supporting the hypothesis that receptor phosphorylation is mediated by the Jaks.

Cytokines couple ligand binding to the tyrosine phosphorylation of a number of proteins that are implicated in signal transduction. Of particular importance is the tyrosine phosphorylation of SHC, the adapter protein that couples the activation of the cytokine receptor superfamily members with activation of the ras pathway. It is hypothesized that SHC is recruited to the receptor complex through recognition of phosphotyrosine "docking" sites by its SH2 domain. In the case of the Epo receptor, the phosphorylation of SHC is dependent upon the membrane distal region of the receptor on which all the detectable phosphorylations occur, consistent with the hypothesis. However, the sites of tyrosine phosphorylation of the receptor have not been identified.

Although most cytokines induce SHC phosphorylation and activation of the ras pathway, IL-4 is notable in that it does neither. However, IL-4 appears to uniquely induce the tyrosine phosphorylation of a protein, termed 4PS, which is functionally and structurally related to insulin response substrate-1 (IRS-1) (Keegan et al., 1994; Wang et al., 1993; Satoh et al., 1991).

10. Regulation of Cytokine Receptor Complexes by Tyrosine Phosphatases

Protein tyrosine phosphatases play a critical role in cytokine signaling, as initially demonstrated by the observation that inhibitors of protein tyrosine phosphatases can partially relieve the requirements for cytokines in mitogenic responses (Tojo et al., 1987). However, their importance became clear in studies of a myeloid-specific enzyme termed hematopoietic cell phosphatase (HCP) (Yi et al., 1992), been termed PTP1C (Shen et al., 1991), SHP (Matthews et al., 1992), and SHPTP1 (Plutzky et al., 1992). HCP is a 68 kDa protein that contains a carboxyl catalytic domain and two SH2 domains in the amino terminal half of the protein. The importance of HCP came with the demonstration that it is the gene responsible for the *motheaten* mutation in mice, indeed, the lethal form of *motheaten* represented a naturally occurring gene knock-out of HCP (Shultz et al., 1993; Tsui et al., 1993). Homozygous *motheaten* mice die within about 3 wk after birth (Van Zant and Shultz, 1989; Shultz, 1991), due largely to overproliferation of macrophages, particularly in the lungs. However, a number of hematopoietic lineages are affected, including the erythroid lineage, in which erythropoiesis becomes relatively independent of Epo. There is also excessive proliferation of the lymphoid lineages, leading to the appearance of autoimmune-like pathologies, possibly by affecting the thresholds for negative selection (Cyster and Goodnow, 1995). This phenotype is consistent with the hypothesis that HCP is required to negatively regulate most lymphoid and myeloid lineages.

HCP is hypothesized to negatively regulate receptor complexes by being recruited to the activated, tyrosine phosphorylated receptor complex through its SH2 domains. In particular, HCP binds, through the amino terminal SH2 domain, to the ligand activated, tyrosine phosphorylated form of c-kit (Yi and Ihle, 1993), the receptor protein tyrosine kinase for stem cell factor (SCF). HCP also binds, through its amino-terminal SH2 domain, to the tyrosine phosphorylated form of the IL3 receptor β chain (Yi *et al.*, 1993) and the tyrosine phosphorylated form of the Epo receptor (Yi *et al.*, 1995). Once recruited to the receptor complex, HCP is positioned to dephosphorylate the activation site of the Jaks and thereby down-regulate the receptor complex.

Tyrosine phosphorylation of the Epo receptor requires the carboxyl-region of the cytoplasmic domain; this is not required for mitogenesis (Miura *et al.*, 1991), but negatively regulates the response to Epo in certain cell lines (D'Andrea *et al.*, 1991). This region also has a profound effect on the function of the receptor *in vivo*. In one form of genetically acquired erythrocytosis, in particular a mutation in the Epo receptor gene results in a 70 amino acid carboxyl-truncation of the receptor (De La Chapelle *et al.*, 1993). Thus, it is hypothesized that both in cell lines and *in vivo* the inability of the Epo receptor to recruit HCP, results in a more active receptor complex. Consistent with this hypothesis, it has been shown that the inability to recruit HCP to the receptor complex is associated with a prolonged activation of Jak2 (Klingmuller *et al.*, 1995).

SHPTP2 (also termed Syp) is a cytoplasmic protein tyrosine kinase which, like HCP, contains an amino terminal region that consists of two SH2 domains and a carboxyl catalytic domain. IL-3, Epo, and GM-CSF induce the tyrosine phosphorylation of SHPTP2 (Welham *et al.*, 1994; Tauchi *et al.*, 1995), which is associated with an increase in phosphatase activity and the acquisition of the ability to associate with Grb2 and the p85 subunit of PI 3'-kinase. IL-6 has also been found to induce the tyrosine phosphorylation of SHPTP2 (Stahl *et al.*, 1995); this is dependent upon a membrane proximal tyrosine that is speculated to be the docking site. The significance of the recruitment of SHPTP2 to the receptor complex is not known. The *Drosophila* homolog of SHPTP2, genetically, is a positive regulator in signaling; it has thus been speculated that SHPTP2 would similarly function in a positive manner in signal transduction.

11. Signal Transducers and Activation of Gene Transcription in Cytokine Signaling

A novel family of transcription factors, termed signal transducers and activators of transcription (Stats), was initially identified in studies concerned with the regulation of gene transcription by the IFNs (Darnell, Jr. *et al.*, 1994). The family currently consists of seven members: Stat1, Stat2, Stat3, Stat4, Stat5a, Stat5b, and

Table II. Properties of the Stat Proteins.

Stat Protein	Chromosome	Optimal Binding Sequence
Stat1	1	TTCC(G>C)GGAA
Stat2	10	None identified
Stat3	11	TTCC(G=C)GGAA
Stat4	1	TTCC(G>C)GGAA
Stat5a	11	TTCC(A>T)GGAA
Stat6	10	TTCCANGGAA

Stat6 (IL4-Stat), which may have evolved from a series of gene duplications. In particular, Stat1 and Stat4 co-localize to murine chromosome 1 (Table II), Stat2 and Stat6 co-localize to chromosome 10, and Stat5a/Stat5b and Stat3 co-localize to chromosome 11 (N. Copeland, N. Jenkins et al., submitted). It should be noted that Stat5a and Stat5b are highly related genes that further co-localize to within at least 100 kb. Thus one can hypothesize that a primordial Stat gene existed and duplicated; this duplicated pair was then duplicated to other chromosomal sites, and finally, recently, another duplication of the Stat5 primordial locus occurred.

The Stat family members are highly related in overall structure. The most characteristic domain is carboxyl-terminal SH2 domain that is required for multiple functions of the Stats. Carboxyl to the SH2 domain is a site of tyrosine phosphorylation, initially identified in Stat1 (Y^{701}) as the critical site that must phosphorylated to induce Stat dimerization and acquisition of DNA binding activity. Each of the Stats contains a comparable site of tyrosine phosphorylation, that is critical for function, comparable to Y^{701} in Stat1. The DNA binding domain of the Stats is novel, and has been recently localized to the middle of the protein (Horvath et al., 1995). This region is highly conserved among the Stats, consistent with the observation that the optimal binding sites for the Stats are closely related, as indicated by binding amplification reactions (Table II). The exception is Stat2, which, while containing the DNA binding motif, does not bind DNA as a homodimer, but is only involved in DNA binding as a heterodimer with Stat1. It should also be noted that the inverted repeat structure of the optimal binding sites is consistent with binding as homo- and heterodimers.

The functional significance of the activation of Stat proteins in various cytokine responses is of considerable interest. Stat1 and Stat2 are initially identified in the context of the regulation of a variety of genes induced the IFNs, many of which are involved in the establishment of an antiviral response (Darnell, Jr. et al., 1994). However, the activation of Stat1 has been seen in a variety of other cytokine responses, as well as in responses to ligands that activate receptor tyrosine kinases (Fu and Zhang, 1993). It is, thus, possible that Stat1 is required for other functions, and consequently that the phenotype of mice in which the Stat 1 gene is disrupted will be of considerable interest. Stat3 was isolated as the transcription factor involved in the IL-6 mediated regulation of a subset of acute phase response proteins (Akira *et al.*, 1994). However, Stat3 is activated by many cytokines and, as with Stat1, may play a much more general role. Stat4 was cloned by homology approaches and, to date, has only been shown to be activated by IL-12 (Jacobson *et al.*, 1994); consequently little is known about genes that may be regulated by Stat4. Stat5 was isolated as a prolactin induced transcription factor mediating the increased transcription of milk proteins in breast cells (Wakao *et al.*, 1994). However, Stat5 is activated in response to a variety of cytokines, and may play quite different roles in the various cell lineages in which it is activated (O. Garra and Murphy, 1994; Perez *et al.*, 1993; Zhang *et al.*, 1994; Suzuki *et al.*, 1995; Muller, 1995; Turunen *et al.*, 1994; Wakao *et al.*, 1995; Azam *et al.*, 1995; Nuchprayoon *et al.*, 1994). Finally, Stat6 was identified as an IL-4 induced transcription factor that binds to a response element of the MHC or the immunoglobulin switch region (Hou *et al.*, 1994; Quelle *et al.*, 1995).

One of the major questions concerning the common activation of Stat proteins in response to cytokines is whether this response is an essential component of a mitogenic response. A variety of observations indicate that this is not the case. First, the "classic" Stat proteins, Stat1 and Stat2, are activated by cytokines that do not induce a mitogenic response but rather, in some cell types, suppress proliferation. Moreover, receptor mutants have been obtained for many cytokine receptors in which the activation of Stat3 (Stahl *et al.*, 1995), Stat5 (Fujii *et al.*, 1995), and Stat6 (Quelle *et al.*, 1995) can be dissociated from the mitogenic responses. Thus it is unlikely that Stat activation by transforming versions of kinases contributes to cellular transformation (Tsarfaty *et al.*, 1994; Lam *et al.*, 1994). However, as with the Jaks, the phenotypes of the mice in which each of the Stat genes is disrupted will be of considerable interest.

12. Conclusions

The last four years have witnessed a tremendous increase in our understanding of the role played by receptors of the cytokine receptor superfamily. This is the result of work from a number of laboratories that have carefully dissected the functional domains of the receptors. Advances have also come from the realization that members of the cytokine receptor superfamily couple ligand binding to the tyrosine phosphorylation and activation of components of signal transduction pathways that have been extensively studied in other receptor systems. Coupling the receptor mutants with the identification of discreet signaling events has made possible the determination of the functional significance of many of the pathways. One of the major advances has clearly been the identification of the once obscure family of cytoplasmic kinases, the Jaks, as central to the function of the cytokine receptor superfamily. Moreover, the identification of the Stat proteins as common elements in signal transduction has provided a richness in responses that was not previously appreciated.

References

Akira, S., Nishio, Y., Inoue, M., Wang, X., Wei, S., Matsusaka, T., Yoshida, K., Sudo, T., Naruto, M., and Kishimoto, T., 1994, Molecular cloning of APRF, a novel ISGF3 p91-related transcription factor involved in the gp130-mediated signaling pathway, *Cell*, **77**:63–71.

Artgetsinger, L.S., Campbell, G.S., Yang, X., Witthuhn, B.A., Silvernnoinen, O., Ihle, J.N., and Carter-Su, C., 1993, Identification of JAK2 as a growth hormone receptor-associated tyrosine kinase, *Cell*, **74**:237–244.

Attisano, L., Wrana, J.L., Lopez-Casillas, F., and Massague, J., 1994, TGF-beta receptors and actions, *Biochim. Biophys. Acta*, **1222**:71–80.

Azam, M., Erdjument-Bromage, H., Kreider, B., Xia, M., Quelle, F.W., Basu, R., Saris, C., Tempst, P., Ihle, J.N., and Schindler, C. 1995, Purification of interleukin-3 stimulated DNA binding factors demonstrates the involvement of multiple isoforms of Stat5 in signaling, *EMBO J.*, **14**(7):1402–1411.

Bacon, C.M., McVicar, D.W., Ortaldo, J.R., Rees, R.C., O'Shea, J.J., and Johnston, J.A., 1995, Interleukin-12 induces tyrosine phosphorylation of JAK2 and TYK2: differential use of Janus tyrosine kinases by interleukin-2 and interleukin-12, *J. Exp. Med.*, **181**:399–404.

Bartley, T.D., Bogenberger, J., Hunt, P., Li, Y.-S., Lu, H.S., Martin, F., Chang, M.-S., Samal, B., Nichol, J.L., Swift, S., Johnson, M.J., Hsu, R.-Y., Parker, V.P., Suggs, S., Skrine, J.D., Merewether, L.A., Clogston, C., Hsu, E., Hokom, M.M., Hornkohl, A., Choi, E., Pangelinan, M., Sun, Y., Mar, V., McNinch, J., Simonet, L;, Jacobson, F., Xie, C., Shutter, J., Chute, H., Basu, R., Selander, L., Trollinger, D., Sieu, L., Padilla, D., Trail, G., Elliott, G., Izumi, R., Covey, T., Crouse, J., Garcia, A., Xu, W., Del Castillo, J., Biron, J., Cole, S., Hu, M.C.-T., Pacifici, R., Ponting, I., Saris, C., Wen, D., Yung, Y.P., Lin, H., and Bosselman, R.A., 1994, Identification and cloning of a megakaryocyte growth and development factor that is a ligand for the cytokine receptor mpl, *Cell*, **77**:1117–1124.

Bazan, J.F., 1990, Structural design and molecular evolution of a cytokine receptor superfamily, *Proc. Natl. Acad. Sci. U.S.A.*, **87**:6934–6938.

Binari, R., and Perrimon, N., 1994, Stripe-specific regulation of pair-rule genes by *hopscotch*, a putative Jak family tyrosine kinase in *Drosophila, Genes & Development*, **8**:300–312.

Brunn, G.J., Falls, E.L., Nilson, A.E., and Abraham, R.T., 1995, Protein-tyrosine kinase-dependent activation of STAT transcription factors in interleukin-2 or interleukin-4-stimulated T lymphocytes, *J. Biol. Chem.*, **270**:11628–11635.

Campbell, G.S., Argetsinger, L.S., Ihle, J.N., Kelly, P.A., Rillema, J.A., and Carter-Su, C., 1994, Activation of JAK2 tyrosine kinase by prolactin receptors in Nb2 cells and mouse mammary gland explants, *Proc. Natl. Acad. Sci. USA*, **91**:5232–5236.

Chua, A.O., Chizzonite, R., Desai, B.B., Truitt, T.P., Nunes, P., Minetti, L.J., Warrier, R.R., Presky, D.H., Levin, J.F., Gately, M.K., and Gubler, U., 1994, Expression cloning of a human IL-12 receptor component: A new member of the cytokine receptor superfamily with strong homology to gp130, *J. Immunology*, **153**:128–136.

Colamonici, O.R., Uyttendaele, H., Domanski, P., Yan, H., and Krolewski, J.J., 1994, p135^{tyk2}, an inteferon α (IFN-α)-activated tyrosine kinase, associates with the IFNα receptor, *J. Biol. Chem.*, **269**:3518–3522.

Corey, S.J, Burkhardt, A.L., Bolen, J.B., Geahlen, R., Tkatch, L.S., and Tweardy, D.J., 1994, Granulocyte colony-stimulating factor receptor signaling involves the formation of a three-component complex with Lyn and Syk protein-tyrosine kinases, *Proc. Natl. Acad. Sci. USA*, **91**:4683–4687.

Cyster, J.G., and Goodnow, C.C., 1995, Protein tyrosine phosphatase 1C negatively regulates antigen receptor signaling in B lymphocytes and determines thresholds for negative selection, *Immunity*, **2**:1–12.

D'Andrea, A.D., Yoshimura, A., Youssoufian, H., Zon, L.I., Koo, J.W., and Lodish, H.F., 1991, The cytoplasmic region of the erythropoietin receptor contains nonoverlapping positive and negative growth-regulatory domains, *Mol. Cell Biol.*, **11**:1980–1987.

Darnell, J.E., Jr., Kerr, I.M., and Stark, G.R., 1994, Jak-STAT pathways and transcriptional activation in response to IFNs and other extracellular signaling proteins, *Science*, **264**:1415–1421.

DaSilva, L., Howard, O.M.Z., Rui, H., Kirken, R.A., and Farrar, W.L., 1994, Growth signaling and JAK2 association mediated by membrane-proximal cytoplasmic regions of prolactin receptors, *J. Biol. Chem.*, **269**:18267–18270.

De La Chapelle, A., Traskelin, A., and Juvonen, E., 1993. Truncated erythropoietin receptor causes dominantly inherited benign human erythrocytosis, *Proc. Natl. Acad. Sci. USA*, **90**:4495–4499.

De Vos, A.M., Ultsch, M., and Kossiakoff, A.A., 1992, Human growth hormone and extracellular domain of its receptor: crystal structure of the complex, *Science*, **255**:306–312.

Dong, F., van Paassen, M., van Buitenen, C., Hoefsloot, L.H., Lowenberg, B., and Touw, I.P., 1995, A point mutation in the granulocyte colony-stimulating factor receptor (G-CSF-R) gene in a case of acute myeloid leukemia results in the overexpression of a novel G-CSF-R isoform, *Blood*, **85**:902–911.

Drachman, J.G., Griffin, J.D., and Kaushansky, K., 1995, The c-mpl ligand (thrombopoietin) stimulates tyrosine phosphorylation of Jak2, Shc, and c-mpl, *J. Biol. Chem.*, **270**(10):4979–4982.

Dusanter-Fourt, I., Muller, O., Ziemiecki, A., Mayeux, P., Drucker, B., Djiane, J., Wilks, A., Harpur, A.G., Fischer, S., and Gisselbrecht, S., 1994, Identification of JAK protein tyrosine kinases as signaling molecules for prolactin. Functional analysis of prolactin receptor and prolactin-erythropoietin receptor chimera expressed in lymphoid cells, *EMBO J.*, **13**:2583–2591.

Ebert, S.N., Balt, S.L., Hunter, J.P.B., Gashler, A., Sukhatme, V., and Wong, D.L., 1994, Egr-1 activation of rat adrenal phenylethanolamine N-methyltransferase gene, *J. Biol. Chem.*, **269**:20885–20898.

Eder, M., Ernst, T.J., Ganser, A., Jubinsky, P.T., Inhorn, R., Hoelzer, D., and Griffin, J.D., 1994, A low affinity human GM-CSF alpha/beta chimeric receptor induces ligand-dependent proliferation in a murine cell line, *J. Biol. Chem.* (in press).

Firmbach-Kraft, I., Byers, M., Shows, T., Dalla-Favera, R., and Krolewski, J.J., 1990, Tyk2, prototype of a novel class of non-receptor tyrosine kinase genes, *Oncogene*, **5**:1329–1336.

Fu, X., and Zhang, J. 1993. Transcription factor p91 interacts with the epidermal growth factor receptor and mediates activation of the c-*fos* gene promoter, *Cell*, **74**:1135–1145.

Fujii, H., Nakagawa, Y., Schindler, U., Kawahara, A., Gouilleux, F., Groner, B., Ihle, J.N., Minami, Y., Miyazaki, T., and Taniguchi, T., 1995, Activation of Stat5 by IL-2 requires a carboxyl terminal region of the IL-2 receptor β chain but is dispensable for the proliferative signal transmission, *Proc. Natl. Acad. Sci. USA* (submitted).

Fukunaga, R., Ishizaka-Ikeda, E., and Nagata, S., 1993, Growth and differentiation signals mediated by different regions in the cytoplasmic domain of granulocyte colony-stimulating factor receptor, *Cell*, **74**:1079–1087.

Giri, J.G., Ahdieh, M., Eisenman, J., Shanebeck, K., Grabstein, K., Kumaki, S., Namen, A., Park, L.S., Cosman, D., and Anderson, D., 1994, Utilization of the beta and gamma chains of the IL-2 receptor by the novel cytokine IL-15, *EMBO J.*, **13**:2822.

Gollob, K.J., and Coffmann, R.L., 1994, A minority subpopulation of CD4[+] T cells directs the development of naive CD4[+] T cells into IL-4-secreting cells, *J. Immunology*, **22**:1–9.

Guschin, D., Rogers, N., Briscoe, J., Witthuhn, B., Watling, D., Horn, F., Pellegrini, S., Yasukawa, K., Heinrich, P., Stark, G.R., Ihle, J.N., and Kerr, I.M., 1995, A major role for the protein tyrosine kinase JAK1 in the JAK/STAT signal transduction pathway in response to interleukin-6, *EMBO J.* (in press).

Hanazono, Y., Chiba, S., Sasaki, K., Mano, H., Miyajima, A., Arai, K., Yazaki, Y., and Hirai, H., 1993a, c-fps/fes protein-tyrosine kinase is implicated in a signalling pathway triggered by granulocyte-macrophage colony-stimulating factor and interleukin-3, *EMBO J.*, **12**:1641–1646.

Hanazono, Y., Chiba, S., Sasaki, K., Mano, H., Yazaki, Y., and Hirai, H., 1993b, Erythropoietin induces tyrosine phosphorylation and kinase activity of the *c-fps/fes* proto-oncogene product in human erythropoietin-responsive cells, *Blood*, **81**:3193–3196.

Hanratty, W.P., and Dearolf, C.R., 1993, The *Drosophila Tumorous-lethal* hematopoietic oncogene is a dominant mutation in the *hopscotch* locus, *Mol. Gen. Genet.*, **238**:33–37.

Ho, A.S., Liu, Y., Khan, T.A., Hsu, D.H., Bazan, J.F., and Moore, K.W., 1993, A receptor for interleukin 10 is related to inteferon receptors, *Proc. Natl. Acad. Sci. USA*, **90**:11267–11271.

Horvath, C.M., Wen, Z., and Darnell, J.E., Jr., 1995, A STAT protein domain which determines DNA sequence recognition suggests a novel DNA binding domain, *Genes & Development* (in press).

Hou, J., Schindler, U., Henzel, W.J., Ho, T.C., Brasseur, M., and KcKnight, S.L., 1994, An interleukin-4-induced transcription factor:IL-4 stat, *Science*, **265**:1701–1706.

Igarashi, K., Garotta, G., Ozmen, L., Ziemiecki, A., Wilks, A.F., Harpur, A.G., Larner, A.C., and Finbloom, D.S., 1994, Interferon-gamma induces tyrosine phosphorylation of interferon-gamma receptor and regulated association of protein tyrosine kinases, Jak1 and Jak2, with its receptor, *J. Biol. Chem.*, **269**:14333–14336.

Ihle, J.N., Witthuhn, B.A., Quelle, F.W., Yamamoto, K., Thierfelder, W.E., Kreider, B., and Silvennoinen, O., 1994, Signaling by the cytokine receptor superfamily: JAKs and STATs, *Trends in Biochemical Sciences*, **19**:222–227.

Ihle, J.N., Witthuhn, B.A., Quelle, F.W., Yamamoto, K., and Silvernnoinen, O., 1995, Signaling through the hematopoietic cytokine receptors, *Annu. Rev. Immunol.*, **13**:369–398.

Jacobson, N.G., Szabo, S., Weber-Nordt, R.M., Zhong, Z., Schreiber, R.D., Darnell, J.E., Jr., and Murphy, K.M., 1994, Interleukin 12 activates Stat3 and Stat4 by tyrosine phosphorylation in T cells, *J. Exp. Med.* (in press).

John, J., McKendry, R., Pellegrini, S., Flavell, D., Kerr, I.M., and Stark, G.R., 1991, Isolation and characterization of a new mutant human cell line unresponsive to alpha and beta interferons, *Mol. Cell Biol.*, **11**:4189–4195.

Johnson, J.A. Kawamura, M., Kirken, R., Chen, Y., Blake, T.B., Shibuya, K., Ortaldo, J.R., McVicar, D.W., and O'Shea, J.J., 1994, Phosphorylation and activation of the JAK3 Janus kinase in response to IL-2, *Nature*, **370**:151–153.

Kakkis, E., Riggs, K.J., Gillespie, W., and Calame, K., 1989, A transcriptional repressor of c-myc, *Nature*, **339**:718–721.

Kaushansky, K., Lok, S., Holly, R.D., Broudy, V.C., Lin, N., Bailey, M.C., Forstrom, J.W., Buddle, M.M., Oort, P.J., and Hagen, F.S., 1994, Promotion of megakaryocyte progenitor expansion and differentiation by the c-Mpl ligand thrombopoietin [*see comments*], *Nature*, **369**(6481):568–571.

Kawamura, M., McVicar, D.W., Johnston, J.A., Blake, T.B., Chen, Y., Lal, B.K., Lloyd, A.R., Kelvin, D.I., Staples, J.E., Ortaldo, J.R., and O'Shea, J.J., 1994, Molecular cloning of L-JAK, a Janus family protein-tyrosine kinase expressed in natural killer cells and activated leukocytes, *Proc. Natl. Acad. Sci. USA*, **91**:6374–6378.

Keegan, A.D., Nelms, K., White, M., Wang, L., Pierce, J.H., and Paul, W.E., 1994, An IL-4 receptor region containing an insulin receptor motif is important for IL-4 mediated IRS-1 phosphorylation and cell growth, *Cell*, **76**:811–820.

Kirken, R.A., Rui, H., Malabraba, M.G., and Farrar, W.L., 1994, Identification of interleukin-2 receptor-associated tyrosine kinase p116 as novel leukocyte-specific Janus kinase, *J. Biol. Chem.*, **269**:19136–19141.

Klingmuller, U., Lorenz, U., Cantley, L.C., Neel, B.G., and Lodish, H.F., 1995, Specific recruitment of the hematopoietic protein tyrosine phosphatase SH-PTP1 to the erythropoietin receptor causes inactivation of JAK2 and termination of proliferative signals, *Cell*, **80**:729–738.

Krosl, J., Damen, J.E., Krystal, G., and Humphries, R.K., 1995, Erythropoietin and interleukin-3 induce distinct events in erythropoietin receptor-expression Ba/F3 cells, *Blood*, **85**:50–56.

Lam, P.Y.S., Jadhav, P.K., Eyermann, C.J., Hodge, C.N., Ru, Y., Bacheler, L.T., Meek, J.L., Otto, M.J., Raynder, M.M., Wong, Y.N., Chang, C., Weber, P.C. Jackson, D.A. Sharpe, T.R., and Erickson-Viitanen, S., 1994, Rational design of potent, bioavailable, nonpeptide cyclic ureas as HIV protease inhibitors, *Science*, **263**:380–384.

Lebrun, J.J., Ali, S., Sofer, L., Ullrich, A., and Kelly, P.A., 1994, Prolactin-induced proliferation of Nb2 cells involves tyrosine phosphorylation of the prolactin receptor and its associated tyrosine kinase JAK2, *J. Biol. Chem.*, **269**:14021–14026.

Leonard, W.J., Noguchi, M., Russell, S.M., and McBride, O.W., 1994, The molecular basis of X-linked severe combined immunodeficiency: the role of the interleukin-2 receptor gamma chain as a common gamma chain, gamma c, *Immunol. Rev.*, **138**:61–86:61–86.

Leung, D.W., Spencer, S.A., Cachianes, G., Hammonds, R.G., Collins, C., Henzel, W.J., Barnard, R., Waters, M.J., and Wood, W.I., 1987, Growth hormone receptor and serum binding protein: purification, cloning and expression, *Nature*, **330**:537–543.

Lew, D.J., Decker, T., and Darnell, Jr., 1989, Alpha interferon and gamma interferon stimulate transcription of a single gene through different signal transduction pathways, *Mol. Cell. Biol.*, **9**:5404–5411.

Luo, H., Hanratty, W.P., and Dearolf, C.R., 1995, An amino acid substitution in the *Drosophila* hop^{Tum-1} Jak kinase causes leukemia-like hematopoietic defects, *EMBO J.* (in press).

Mano, H., Yamashita, Y., Sato, K., Yazaki, Y., and Hirai, H., 1995, Tec protein-tyrosine kinase is involved in interleukin-3 signaling pathway, *Blood*, **85**:343–350.

Matsuda, T., Fukada, T., Takahashi-Tezuka, M., Okuyama, Y., Fujitani, Y., Hanazono, Y., Hirai, H., and Hirano, T., 1995a.Activation of fes tyrosine kinase by gp130, an interleukin-6 family cytokine signal transducer, and their association, *J. Biol. Chem.*, **270**:11037–11039.

Matsuda, T., Takahashi-Tezuka, M., Fukada, T., Okuyama, Y., Fujitani, Y., Tsukada, S., Mano, H., Hirai, H., Witte, O.N., and Hirano, T., 1995b. Association and activation of Btk and Tec tyrosine kinases by gp130, a signal transducer of the interleukin-6 family of cytokines, *Blood*, **85**:627–633.

Matthews, R.J., Bowne, D.B., Flores, E., and Thomas, M.L., 1992, Characterization of hematopoietic intracellular protein tyrosine phosphatases: description of a phosphatase containing an SH2 domain and another enriched in proline-, glutamic acid-, serine-, and threonine-rich sequences, *Mol. Cell Biol.*, **12**:2396–2405.

McKendry, R., John, J., Flavell, D., Muller, M., Kerr, I.M., and Stark, G.R., 1991, High-frequency mutagenesis of human cells and characterization of a mutant unresponsive to both α and -γ interferons, *Proc. Natl. Acad. Sci. USA*, **88**:11455–11459.

Minami, Y., Kono, T., Miyaki, T., and Taniguchi, T., 1993, The IL-2 receptor complex: Its structure, function and target genes, *Annu. Rev. Immunol.*, **11**:245–267.

Minami, Y., Nakagawa, Y., Kawahara, A., Miyazaki, T., Sada, K., Yamamura, H., and Taniguischi, T., 1995, Protein tyrosine kinase Syk is associated with and activated by the IL-2 receptor: possible link with the c-myc induction pathway, *Immunity*, **2**:89–100.

Miura, O., D'Andrea, A., Kabat, D., and Ihle, J.N., 1991, Induction of tyrosine phosphorylation by the erythropoietin receptor correlates with mitogenesis, *Mol. Cell. Biol.*, **11**:4895–4902.

Muira, O., Nakamura, N., Quelle, F.W., Witthuhn, B.A., Ihle, J.N., and Aoki, N. 1994, Erythropoietin induces association of the JAK2 protein tyrosine kinase with the erythropoietin receptor *in vivo*, *Blood*, **84**:1501–1507.

Miyajima, A., Mui, A.L.-F., Ogorochi, T., and Sakamaki, K., 1993, Receptors for granulocyte-macrophage colony-stimulating factor, interleukin-3, and interleukin-5, *Blood*, **82**:1960–1974.

Miyazaki, T., Kawahara, A., Minami, Y., Liu, Z., Silvennoinen, O., Witthuhn, B.A., Ihle, J.N., and Taniguchi, T., 1994, IL-2 signaling; Functional activation of Jak1 and Jak3 kinases through selective association with the IL-2R beta and gamma chains, *Science*, **266**:1045–1047.

Muller, M., Briscoc, J., Laxton, C., Guschin, D., Ziemiecki, A., Silvennoinen, O., Harpur, A.G., Barbieri, G., Witthuhn, B.A., Schindler, C., Pellegrini, S., Wilks, A.F., Ihle, J.N., Stark, G.R., and Kerr, I.M., 1993, The protein tyrosine kinase JAK1 complements defects in interferon-$\alpha\beta$ and – signal transduction, *Nature*, **366**:129–135.

Muller, R., 1995, Transcriptional regulation during the mammalian cell cycle, *TIG*, **11**:173–178.

Murakami, M., Hibi, M., Nakagawa, N., Nakagawa, T., Yasukawa, K., Yamanishi, K., Taga, T., and Kishimoto, T., 1993, IL-6-induced homodimerization of gp130 and associated activation of a tyrosine kinase, *Science*, **260**:1808–1810.

Nabavi, N., Grusby, M.J., Finn, P.W., Wolgemuth, D.J., and Glimcher, L.H., 1990, Identification of an IL-4-inducible gene expressed in differentiating lymphocytes and male germ cells, *Developmental Immunology*, **1**:19–30.

Narazaki, M., Witthuhn, B.A., Yoshida, K., Silvennoinen, O., Yasukawa, K., Ihle, J.N., Kishimoto, T., and Taga, T., 1994, Activation of JAK2 kinase mediated by the IL-6 signal transducer, gp130, *Proc. Natl. Acad. Sci. USA*, **91**:2285–2289.

Nelson, B.H., Lord, J.D., and Greenberg, P.D., 1994, Cytoplasmic domains of the interleukin-2 receptor beta and gamma chains mediate the signal for T-cell proliferation, *Nature*, **369**:333–336.

Nicholson, S.E., Oates, A.C., Harpur, A.G., Ziemiecki, A., Wilks, A.F., and Layton, J.E., 1994, Tyrosine kinase JAK1 is associated with the granulocyte-colony-stimulating factor receptor and both become tyrosine-phosphorylated after receptor activation, *Proc. Natl. Acad. Sci. USA*, **91**:2985–2988.

Noguchi, M., Nakamura, Y., Russell, S.M., Ziegler, S.F., Tsang, M., Cao, X., and Leonard, W.J., 1993, Interleukin-2 receptor gamma chain: a functional component of the interleukin-7 receptor [*see comments*], *Science*, **262**(5141):1877–1880.

Novick, D., Cohen, B., and Rubinstein, M., 1994, The human interferon α/β receptor: characterization and molecular cloning, *Cell*, **77**:391–400.

Nuchprayoon, I., Meyers, S., Scott, L.M., Suzow, J., Hiebert, S., and Friedman, A.D., 1994, PEBP2/CBF, the murine homology of the human myeloid AML1 and PEBP2β/CBFβ proto-oncoproteins, regulates the murine myeloperoxidase and neutrophil elastase genes in immature myeloid cells, *Mol. Cell. Biol.*, **14**:5558–5568.

O'Garra, A., and Murphy, K., 1994, Role of cytokines in determining T-lymphocyte function, *Current Biology*, **6**:458–466.

Obiri, N.I., Debinski, W., Leonard, W.J., and Puri, R.K., 1995, Receptor for interleukin 13 Interaction with interleukin 4 by a mechanism that does not involve the common gamma chain shared by receptors for interleukin 2, 4, 7, 9 and 15, *J. Biol. Chem.*, **270**(15):8797–8804.

Pellegrini, S., John, J., Shearer, M., Kerr, I.M., and Stark, G.R., 1989, Use of a selectable marker regulated by alpha interferon to obtain mutations in the signalling pathway, *Mol. Cell. Biol.*, **9**:4605–4612.

Perez, C., Wietzerbin, J., and Benech, P.D., 1993, Two *cis*-DNA elements involved in myeloid-cell-specific expression and gamma interferon (IFN-γ) activation of the human high-affinity Fcγ receptor gene: a novel IFN regulatory mechanism, *Mol. Cell. Biol.*, **13**:2182–2192.

Plutzky, J., Neel, B.G., and Rosenberg, R.D., 1992, Isolation of a src homology 2-containing tyrosine phosphatase, *Proc. Natl. Acad. Sci. USA*, **89**:1123–1127.

Pritchard, M.A., Baker, E., Callen, D.F., Sutherland, G.R., and Wilks, A.F., 1992, Two members of the JAK family of protein tyrosine kinases map to chromosomes 1p31.3 and 9p24, *Mamm. Genome*, **3**:36–38.

Quelle, F.W., Sato, N., Witthuhn, B.A., Inhorn, R., Ernst, T.J., Miyajima, A., Griffin, J.D., and Ihle, J.N., 1994, JAK2 associates with the β_c chain of the receptor for GM-CSF and its activation requires the membrane proximal region, *Mol. Cell. Biol.*, **14**:4335–4341.

Quelle, F.W., Shimoda, K., Thierfelder, W., Fischer, C., Kim, A., Ruben, S.M., Cleveland, J.L., Pierce, J.H., Keegan, A.D., Nelms, K., Paul, W.E., and Ihle, J.N., 1995, Cloning of murine and human Stat6 (IL-4-Stat): A novel stat tyrosine phosphorylated in the responses to IL-4 and IL-3 that is not required for mitogenesis, *Mol. Cell. Biol.* in press.

Rane, S.G., and Reddy, E.P., 1994, JAK3: A novel JAK kinase associated with terminal differentiation of hematopoietic cells, *Oncogene*, **9**:2415–2423.

Rawlings, D.J., Saffran, D.C., Tsukada, S., Largaespada, D.A., Grimaldi, J.C., Cohen, L., Mohr, R.N., Bazan, J.F., Howard, M., Copeland, N.G., Jenkins, N.A., and Witte, O.N., 1993, Mutation of unique region of Bruton's tyrosine kinase in immunodeficient XID mice, *Science*, **261**:358–361.

Renauld, J.C., Druez, C., Kermouni, A., Houssiau, F., Uyttenhove, C., Van Roost, E., and Van Snick, J., 1992, Expression cloning of the murine and human interleukin 9 receptor cDNAs, *Proc. Natl. Acad. Sci. USA*, **89**(12):5690–5694.

Rivera, V.M., and Brugge, J.S., 1995, Clustering of Syk is sufficient to induce tyrosine phosphorylation and release of allergic mediators from rat basophilic leukemia cells, *Mol. Cell. Biol.*, **15**(3):1582–1590.

Rolling, C., Treton, C., Beckmann, P., Galanaud, P., and Richard, Y., 1995, JAK3 associates with the human interleukin 4 receptor and is tyrosine phosphorylated following receptor triggering, *Oncogene*, **10**:1757–1761.

Rui, H., Kirken, R.A., and Farrar, W.L., 1994, Activation of receptor-associated tyrosine kinase JAK2 by prolactin, *J. Biol. Chem.*, **269**:5364–5368.

Russell, S.M., Johnston, J.A., Noguchi, M., Kawamura, M., Bacon, C.M., Friedmann, M., Berg, M., McVicar, D.W., Witthuhn, B.A., Silvennoinen, O., Goldman, A.S., Schmalstieg, F.C., Ihle, J.N., O'Shea, J.J., and Leonard, W.J., 1994, Interaction of IL-2 receptor β and γ_c chains with JAK1 and JAK3, respectively: Defective γ_c-JAK3 association in XSCID, *Science*, **266**:1042–1045.

Sato, S., Katagiri, T., Takaki, S., Kikuchi, Y., Hitoshi, Y., Yonehara, S., Tsukada, S., Kitamura, D., Watanabe, T., Witte, O. *et al.*, 1994b, IL-5 receptor-mediated tyrosine phosphorylation of SH2/SH3-containing proteins and activation of Bruton's tyrosine and Janus 2 kinases. *J. Exp. Med.*, **180**:2101–2111.

Satoh, T., Nakafuku, M., Miyajima, A., and Kaziro, Y., 1991, Involvement of ras p21 protein in signal-transduction pathways from interleukin 2, interleukin 3, and granulocyte/macrophage colony-stimulating factor, but not from interleukin 4, *Proc. Natl. Acad. Sci. USA*, **88**:3314–3318.

Seckinger, P., and Fougereau, M., 1994, Activation of src family kinases in human pre-B cells by IL-7, *J. Immunol.*, **153**(1):97–109.

Shen, S.H., Bastien, L., Posner, B.I., and Chrétien, P., 1991, A protein-tyrosine phosphatase with sequence similarity to the SH2 domain of the protein-tyrosine kinases, *Nature*, **352**:736–739.

Shimoda, K., Iwasaki, H., Okamura, S., Ohno, Y., Kubota, A., Arima, F., Otsuka, T., and Niho, Y., 1994, G-CSF induces tyrosine phosphorylation of the Jak2 protein in the human myeloid G-CSF responsive and proliferative cells, but not in mature neutrophiles, *Biochem. Biophys. Res. Commun.*, **203**:922–928.

Shultz, L.D., 1991, Hematopoiesis and models of immunodeficiency, *Sem. Immunol.*, **3**:397–408.

Shultz, L.D., Schweitzer, P.A., Rajan, T.V., Yi, T., Ihle, J.N., Matthews, R.J., Thomas, M.L., and Beier, D.R., 1993, Mutations at the murine motheaten locus are within the hematopoietic cell protein tyrosine phosphatase (Hcph) gene, *Cell*, **73**:1445–1454.

Silvennoinen, O., Witthuhn, B., Quelle, F.W., Cleveland, J.L., Yi, T., and Ihle, J.N., 1993, Structure of the JAK2 protein tyrosine kinase and its role in IL-3 signal transduction, *Proc. Natl. Acad. Sci. USA*, **90**:8429–8433.

Soh, J., Donnelly, R.J., Kotenko, S., Mariano, T.M., Cook, J.R., Wang, N., Emmanuel, S., Schwartz, B., Miki, T., and Pestka, S., 1994, Identification and sequence of an accessory factor required for activation of the human interferon gamma receptor, *Cell*, **76**:793–802.

Somers, W., Ultsch, M., De Vos, A.M., and Kossiakoff, A.A., 1994, The X-ray structure of a growth hormone-prolactin receptor complex, *Nature*, **372**:478–481.

Stahl, N., and Yancopoulos, G.D., 1993, The alphas, betas and kinases of cytokine receptor complexes, *Cell*, **74**:587–590.

Stahl, N., Boulton, T.G., Farruggella, T., Ip, N.Y., Davis, S., Witthuhn, B.A., Quelle, F.W., Silvennoinen, O., Barbieri, G., Pellegrini, S., Ihle, J.N., and Yancopoulos, G.D., 1994, Association and activation of Jak-Tyk kinases by CNTF-LIF-OSM-IL-6 beta receptor components. *Science*, **263**:92–95.

Stahl, N., Farruggella, T.J., Boulton, T.G., Zhong, Z., Darnell, J.E., Jr., and Yancoupoulos, G.D., 1995, Modular tyrosine-based motifis in cytokine receptors specify choice of stats and other substrates, *Science*, **267**(3 March):1349–1353.

Suzuki, H., Kundig, T.J., Furlonger, C., Wakeham, A., Timms, E., Matsuyama, T., Schmits, R., Simard, J.J.L., Ohashi, P.S., Griesser, H., Tanaguchi, T., Paige, C.J., and Mak, T.W., 1995, Deregulated T cell activation and autoimmunity in mice lacking interleukin-2 receptor β, *Science*, **268**:1472–1476.

Tan, J.C., Indelicato, S.R., Narula, S.K., Zavodny, P.J., and Chou, C.C., 1993, Characterization of interleukin-10 receptors on human and mouse cells, *J. Biol. Chem.*, **268**(28):21053–21059.

Tanaka, N., Asao, H., Ohbo, K., Ishii, N., Takeshita, T., Nakamura, M., Sasaki, H., and Sugamura, K., 1994, Physical association of Jak1 and Jak2 tyrosine kinases with the interleukin 2 receptor β and gamma chains, *Proc. Natl. Acad. Sci. USA*, **91**:7271–7275.

Tang, B., Mano, H., Yi, T., and Ihle, J.N., 1994, Tec kinase associates with c-Kit and is tyrosine phosphorylated and activated following stem cell factor binding, *Mol. Cell. Biol.*, **14**:8432–8437.

Taniguchi, T., 1995, Cytokine signaling through nonreceptor protein tyrosine kinases, *Science*, **268**(14 April):251–255.

Taniguchi, T., and Minami, Y., 1993, The IL-2/IL-2 receptor system: A current overview, *Cell*, **73**:5–8.

Tauchi, T., Feng, G., Shen, R., Hoatlin, M., Bagby, G.C.J., Kabat, D., Lu, L., and Broxmeyer, H.E., 1995, Involvement of SH2-containing phosphotyrosine phosphatase Syp in erythropoietin receptor signal transduction pathways, *J. Biol. Chem.*, **270**(10):5631–5635.

Thomas, J.D., Sideras, P., Smith, C.I., Vorechovsky, I., Chapman, V., and Paul, W.E., 1993, Colocalization of X-linked agammaglobulinemia and X-linked immunodeficiency genes, *Science*, **261**:355–358.

Tojo, A., Kasuga, M., Urabe, A., and Takaku, F., 1987, Vanadate can replace interleukin 3 for transient growth of factor-dependent cells, *Exp. Cell. Res.*, **171**:16–23.

Torigoe, T., O'Connor, R., Santoli, D., and Reed, J.C., 1992, Interleukin-3 regulates the activity of the LYN protein-tyrosine kinase in myeloid-committed leukemic cell lines, *Blood*, **80**:617–624.

Tsarfaty, I., Rong, S., Resau, J.H., Rulong, S., da Silva, P.P., and Vande Woude, G.F., 1994, The met proto-oncogene mesenchymal to epithelial cell conversion, *Science*, **263**:98–101.

Tsui, H.W., Siminovitch, K.A., de Souza, L., and Tsui, F.W.L., 1993, Motheaten and viable motheaten mice have mutations in the haematopoietic cell phosphatase gene, *Nat. Gen.*, **4**:124–129.

Tsukada, S., Saffran, D.C., Rawlings, D.J., Parolini, O., Allen, R.C., Klisak, I., Kubagawa, H., Mohandas, T., Quan, S., Belmont, J.W., Cooper, M.D., Conley, M.E., and Witte, O.N., 1993, Deficient expression of a B-cell cytoplasmic tyrosine kinase in human X-linked agammaglobulinemia, *Cell*, **72**:279–290.

Turunen, O., Wahlstrom, T., and Vaheri, A., 1994, Ezrin has a COOH-terminal actin-binding site that is conserved in the ezrin protein family. *J. Cell. Biol.*, **126**:1445–1453.

Van Zant, G., and Shultz, L., 1989, Hematopoietic abnormalities of the immunodeficient mouse mutant, viable motheaten (*mev*), *Exp. Hematol.*, **17**:81-87.

VanderKuur, J.A., Wang, X., Zhang, L., Campbell, G.S., Allevato, G., Billestrup, N., Norstedt, G., and Carter-Su, C., 1994, Domains of the growth hormone receptor required for association and activation of JAK2 tyrosine kinase, *J. Biol. Chem.*, **269**:21709–21717.

Velazquez, L., Fellous, M., Stark, G.R., and Pellegrini, S., 1992, A protein tyrosine kinase in the interferon α/β signaling pathway, *Cell*, **70**:313–322.

Velazquez, L., Mogensen, K.E., Barbieri, G., Fellous, M., Uze, G., and Pellegrini, S., 1995, Distinct domains of the protein tyrosine kinase tyk2 required for binding of Interferon-α/β and for signal transduction. *J. Biol. Chem.*, **270**(7):3327–3334.

Venkitaraman, A.R., and Cowling, R.J., 1992, Interleukin-7 receptor functions by recruiting the tyrosine kinase p59[fyn] through a segment of its cytoplasmic tail, *Proc. Natl. Acad. Sci. USA*, **89**:12083–12087.

Vogel, L.B., and Fujita, D.J., 1995, p[70] phosphorylation and binding to p56[lck] is an early event in interleukin-2-induced onset of cell cycle progression in T-lymphocytes, *J. Biol. Chem.*, **270**:2506–2511.

Wakao, H., Gouilleux, F., and Groner, B., 1994, Mammary gland factor (MGF) is a novel member of the cytokine regulated transcription factor gene family and confers the prolactin response, *EMBO J*, **13**:2182–2191.

Wakao, H., Harada, N., Kitamura, T., Mui, A.L.-F. and Miyajima, A., 1995, Interleukin 2 and erthropoietin activate STAT5/MGF via distinct pathways, *EMBO J* (in press).

Wang, L., Myers, M.G., Jr. Sun, X., Aaronson, S.A., White, M., and Pierce, J.H., 1993, Expression of IRS-1 restores insulin- and IL-4-mediated mitogenesis in hematopoietic cells, *Science*, **261**:1591–1594.

Watling, D., Guschin, D., Muller, M., Silvennoinen, O., Witthuhn, B.A., Quelle, F.W., Rogers, N.C., Schindler, C., Stark, G.R., Ihle, J.N., and Kerr, I.M., 1993, Complementation by the protein tyrosine kinase JAK2 of a mutant cell line defective in the interferon-τ signal transduction pathway, *Nature*, **366**:166–170.

Watowich, S.S., Hilton, D.J., and Lodish, H.F., 1994, Activation and inhibition of erythropoietin receptor function: role of receptor dimerization, *Mol. Cell. Biol.*, **14**: 3535–3549.

Wei, X., Charles, I.G., Smith, A., Ure, J., Feng, G., Huang, F., Xu, D., Muller, W., Moncada, S., and Liew, F.Y., 1995, Altered immune responses in mice lacking inducible nitric oxide synthase, *Nature*, **375**:408–411.

Weiss, A., and Littman, D.R., 1994, Signal transduction by lymphocyte antigen receptors, *Cell*, **76**:263–274.

Weiss, M., Yokoyama, C., Shikama, Y., Naugle, C., Druker, B., and Sieff, C.A., 1993, Human granulocyte-macrophage colony-stimulating factor receptor signal transduction requires the proximal cytoplasmic domains of the α and β subunits, *Blood*, **82**:3298–3306.

Welham, M.J., Dechert, U., Leslie, K.B., Jirik, F., and Schrader, J.W., 1994, Interleukin (IL)-3 and granulocyte/macrophage colony-stimulating factor, but not IL-4, induce tyrosine phosphorylation, activation, and association of SHPTP2 with Grb2 and phosphatidylinositol 3'-kinase, *J. Biol. Chem.*, **269**:23764–23768.

Witthuhn, B., Quelle, F.W., Silvennoinen, O., Yi, T., Tang, B., Miura, O., and Ihle, J.N., 1993, JAK2 associates with the erythropoietin receptor and is tyrosine phosphorylated and activated following EPO stimulation, *Cell*, **74**:227–236.

Witthuhn, B.A., Silvennoinen, O., Miura, O., Lai, K.S., Cwik, C., Liu, E.T., and Ihle, J.N., 1994, Involvement of the JAK3 Janus kinase in IL-2 and IL-4 signalling in lymphoid and myeloid cells, *Nature*, **370**:153–157.

Yi, T., and Ihle, J.N., 1993, Association of hematopoietic cell phosphatase with c-kit after stimulation with c-kit ligand, *Mol. Cell. Biol.*, **13**:3350–3358.

Yi, T., Cleveland, J.L., and Ihle, J.N., 1992, Protein tyrosine phosphatase containing SH2 domains: characterization, preferential expression in hematopoietic cells, and localization to human chromosome 12p12–p13, *Mol. Cell. Biol.*, **12**:836–846.

Yi, T., Mui, A.L. -F., Krystal, G., and Ihle, J.N., 1993, Hematopoietic cell phosphatase associates with the interleukin-3 (IL-3) receptor β chain and down-regulates IL-3-induced tyrosine phosphorylation and mitogenesis, *Mol. Cell. Biol.*, **13**:7577–7586.

Yi, T., Zhang, J., Miura, O., and Ihle, J.N., 1995, Hematopoietic cell phosphatase (HCP) associates with the erythropoietin receptor following Epo induced receptor tyrosine phosphorylation: Identification of potential binding sites, *Blood*, **85**:87–95.

Yin, T., Tsang, M.L., and Yang, Y., 1994, Jak1 kinase forms complexes with interleukin-4 receptor and 4PS/insulin receptor substrate-1-like protein and is activated by interleukin-4 and interleukin-9 in T lymphocytes, *J. Biol. Chem.*, **269**:26614–26617.

Youssoufian, H., Longmore, G., Neumann, D., Yoshimura, A., and Lodish, H.F., 1993, Structure, function, and activation of the erythropoietin receptor, *Blood*, **9**:2223–2236.

Zeng, Y., Takahashi, H., Shibata, M., and Hirokawa, K., 1994, JAK3 janus kinase is involved in interleukin 7 signal pathway, *FEBS Letters*, **353**:289–293.

Zhang, Z., Maclean, D., McNamara, D.J., Sawyer, T.K., and Dixon, J.E., 1994, Protein tyrosine phosphatase substrate specificity: size and phosphotyrosine positioning requirements in peptide substrates, *Biochemistry*, **33**:2285–2290.

Zhuang, H., Patel, S.V., He, T., Sonsteby, S.K., Niu, A., and Wojchowski, D.M., 1994, Inhibition of erythropoietin-induced mitogenesis by a kinase-deficient form of Jak2, *J. Biol. Chem.*, **269**:21411–21414.

Ziemiecki, A., Harpur, A.G., and Wilks, A.F., 1994, JAK protein tyrosine kinases: their role in cytokine signalling. *Trends in Cell Biology*, **4**:207–212.

Zurawski, S.M., Vega, F.J., Huyghe, B., and Zurawski, G., 1993, Receptors for interleukin-13 and interleukin-4 are complex and share a novel component that functions in signal transduction, *EMBO J.*, **12**:2663–2670.

7

Cytokine Driven Signal Transmission

GERALD A. EVANS, ROY J. DUHE, O.M. ZACK HOWARD,
ROBERT A. KIRKEN, LUIS DA SILVA, REBECCA ERWIN,
MARIA G. MALABARBA AND WILLIAM L. FARRAR

1. The Cytokine Receptor Superfamily

Immune system function is both positively and negatively regulated at the level of cell viability or cycle progression by cytokines whose receptors form a distinct family. This family has been variably referred to as the cytokine/hematopoietin receptor superfamily (reviewed by Cosman, 1993; Bazan, 1990) and more recently as the prolactin/growth hormone/interleukin (PRL/GH/IL) receptor family (Kelly *et al.*, 1991; Kirken *et al.*, 1994). This growing family is composed of the receptors for interleukin-2 (IL-2) through IL-7, IL-9, IL-11, IL-12, IL-13, IL-15, prolactin (PRL), growth hormone (GH), granulocyte colony stimulating factor (G-CSF), granulocyte-macrophage colony stimulating factor (GM-CSF), erythropoietin (EPO), oncostatin M (OSM), leukemia inhibitory factor (LIF), ciliary neurotrophic factor (CNTF), and thrombopoietin (MPL) (Cosman, 1993; Kelly *et al.*, 1991; Mott and Campbell, 1995; Vignon *et al.*, 1992; Vita *et al.*, 1995; Giri *et al.*, 1995).

The cytokine receptor family is structurally characterized by an approximately 200 amino acid conserved extracellular domain containing four positionally conserved cysteine residues and a conserved WSXWS motif found near the

Correspondence: Dr. Gerald A. Evans, BCDP, NCI-FCRDC, Frederick, MD 21702–1201, Tel: (301) 846-1505; Fax: (301) 846-5126.

The content of this publication does not necessarily reflect the views or policies of the Department of Health and Human Services, nor does mention of trade names, commercial products, or organizations imply endorsement by the U.S. Government.

extracellular side of the 25 amino acid transmembrane domain (Bazan, 1990). Comparative analysis of the cytoplasmic domains has revealed that these receptors contain loosely shared regions of homology denoted as homology Box 1 and 2, found adjacent to the transmembrane domain.

Within the hematopoietin receptor superfamily, some receptor complexes can be found to contain only one ligand binding molecule, such as the receptor systems for PRL, GH, EPO, and G-CSF, while others contain two or more distinct receptor molecules that form complexes with a shared receptor molecule; for example, the IL-3, GM-CSF, and IL-5 receptors, which interact with the β_c subunit, the IL-6, LIF, CNTF, and IL-11 receptors, which interact with gp130, or the IL-2 α and β chains, IL-4, and IL-7 receptors, which interact with the γ_c chain (reviewed by Heldin, 1995).

It is clear that receptors within this family contain structural similarities in the extracellular and cytoplasmic domain, and examples can be found of a signal transduction mechanism that utilizes a sharing of receptor molecules. This suggests a common mechanism of signal transduction within the hematopoietin receptor superfamily. Therefore, in order to simplify the discussion of mitogenic signal transduction by these receptors, the focus will be placed on analyzing signaling mechanisms employed by the IL-2 and PRL receptor systems (structurally represented in Fig. 1) as representative hematopoietin receptor models.

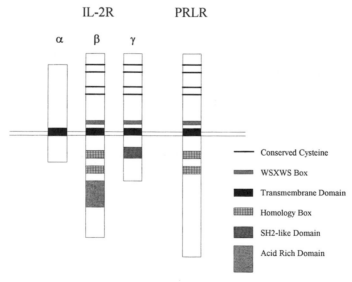

Figure 1. The IL-2 and PRL receptors typify cytokine receptor superfamily members. Conserved extracellular cysteine residues, a conserved WSXWS box in the extracellular domain, and conserved Homology Box domains (Box 1, Box 2) in the proximal cytoplasmic domain define this family. The IL-2 and PRL receptor complexes are examples of this every-growing family, which exemplifies homo- and hetero-dimerization as well as receptor complex formation with non-hematopoietin receptor molecules (IL-2Rα chain).

2. Receptor Aggregation and Tyrosine Kinase Activation Initiates Signal Transduction

Several pieces of evidence establish that the initiation of cytokine signaling involves cytokine binding and dimerization or oligomerization of receptor molecules. Analysis of GH and GH receptor crystallography data reveals that while GH is a monomeric molecule with no apparent symmetry, it is capable of interacting with two GH receptor molecules simultaneously (Cunningham *et al.*, 1991; de Vos *et al.*, 1992; Ultsch *et al.*, 1991). Antibodies specific for the extracellular domain of the PRL receptor have been shown to induce mitogenesis and tyrosine kinase activation in a manner similar to PRL by artificially aggregating monomeric PRL receptors (Rui *et al.*, 1994a). Moreover, monovalent anti-PRL receptor Fab fragments fail to induce cell growth; however, crosslinking receptors with further incubation with bivalent anti-Fab antibodies restores signal transduction and mitogenesis (Rui *et al.*, 1994a).

Both GH and PRL receptor systems contain unique monomeric receptors that associate with, and are dimerized by, cytokine. In contrast, the IL-2 receptor system is composed of heteromeric receptor molecules which include an approximately 75 kDa β chain, a 64kDa γ chain (γ_c) that is shared by a number of cytokine receptors (Kimura *et al.*, 1995; Kondo *et al.*, 1993), and an affinity converting α chain of 55 kDa. All receptor molecules can interact with IL-2 in an activated complex; however, it appears that only the β and γ chains are essential for signal propagation and mitogenesis.

The direct requirement for a β/γ heterodimer in IL-2 signaling can be found amidst data generated from chimeric receptor studies. IL-2 receptor chimeras containing the cytoplasmic domain of the β and γ chains fused to the extracellular cytokine binding domain of the EPO receptor (EPOβ and EPOγ) were generated, and the ability of EPO to bind to these chimeric receptors and induce IL-2 analogous signal was investigated (Goldsmith *et al.*, 1995). Transfection of the human T cell line HT-2 with these chimeric receptors followed by EPO stimulation revealed that neither EPOβ nor EPOγ alone were capable of stimulating an IL-2 response, while co-expression of EPOβ and EPOγ resulted in EPO-dependent proliferation (Goldsmith *et al.*, 1995). The requirement of a β/γ heterodimer for IL-2 signal propagation was also confirmed by similar experiments performed in murine Ba/F3 cells using receptor chimeras consisting of the extracellular domain of c-kit and the IL-2 receptor β and γ cytoplasmic domains (Nelson *et al.*, 1994) and the use of CD4/$\beta\gamma$ chimeras (Miyazaki *et al.*, 1994). It is thus clear that receptor aggregation mediated by cytokine binding to the extracellular domain initiates receptor level signal transduction, and that the quality of this signal is defined by structural motifs within the receptor cytoplasmic domain.

Although cytokine receptors lack any identifiable catalytic domain within the cytoplasmic portion of the receptor, they have all been shown to rapidly induce tyrosine kinase activity. The Janus family of tyrosine kinases (JAKs) have recently been found to be associated with cytokine receptors and to be activated upon cytokine receptor engagement (reviewed by Wilkes and Harper, 1994; Ihle *et al.*, 1994). This is exemplified by the observation that JAK2 is rapidly tyrosine phosphorylated and activated following PRL treatment of rat Nb2 cells (Rui *et al.*, 1994b). Furthermore, this activation has been shown to be the result of a specific association of JAK2 with the PRL receptor (Rui *et al.*, 1994b), and to be dependent upon PRL receptor dimerization (Rui *et al.*, 1994).

Based on chimeric receptor data and signal transduction studies, it is well established that the cytoplasmic domain of the cytokine receptors governs the quality of signal induced by ligand. In order to more fully explore the function of cytoplasmic domain motifs, various receptor deletion mutants of the rat PRL receptor were generated and analyzed for their ability to support JAK activation and mitogenesis. As mentioned, the PRL receptor complex consists of a single polypeptide that forms a homodimer upon PRL interaction (Rui *et al.*, 1994a). As with hematopoietin receptor superfamily receptors, the PRL receptor contains a large extracellular domain and a 356 amino acid cytoplasmic domain containing the membrane proximal Box 1 and Box 2 homology domains (Cosman, 1993). Deletion mutants of this receptor were generated and transfected into murine 32D cells (Fig. 2A). Transfectants were treated with physiological concentrations of PRL and assayed for their ability to support induction of ornithine decarboxylase (ODC) gene transcription, JAK2 association and activation, and mitogenesis (DaSilva *et al.*, 1994).

Several conclusions can be drawn from the results of these experiments. It is clear that a large portion of the cytoplasmic domain of the PRL receptor is not necessary for the propagation of growth signal. This is evident because of the ability of receptor mutant G328, which lacks 262 amino acids of the distal cytoplasmic domain, to support PRL-dependent cell growth (Fig. 2B). Additionally, PRL-dependent activation of JAK2, as measured by anti-phosphotyrosine immunoblotting of JAK2 immune precipitates, and JAK2 association with the PRL receptor, is lost upon deletion of the region spanning the Box 2 homology region (Fig. 3, A and B). This results in an inability of these receptors to induce ODC message and cell growth (Fig. 2, B and C). Altogether, these experiments establish that the Box 2 homology domain of the PRL receptor is required for coupling to PRL-dependent JAK2 activation and receptor association, and that JAK2 activation is required for the propagation of mitogenic signal (DaSilva *et al.*, 1994).

In contrast to the monomeric PRL receptor system, signaling competent IL-2 and IL-4 receptor complexes are composed of at least two distinct receptors that signal via the coordinate activation of multiple JAK kinases. IL-2 and IL-4 have been shown to induce the activity of JAK1 and JAK3 (Johnston *et al.*, 1994; Witthuhn *et al.*, 1994); JAK1 has been found associated with the IL-2 receptor β chain and JAK3 with the γ_c chain (Russell *et al.*, 1994; Miyazaki *et al.*, 1994). As with the PRL receptor, IL-2 and IL-4 receptor mutagenesis studies also revealed a similar requirement for membrane proximal regions of the IL-4 receptor α,

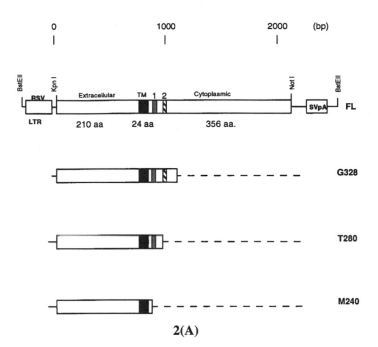

2(A)

PRL-Dependent Proliferation of PRLR Mutants

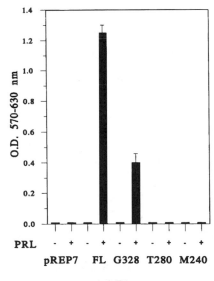

2(B)

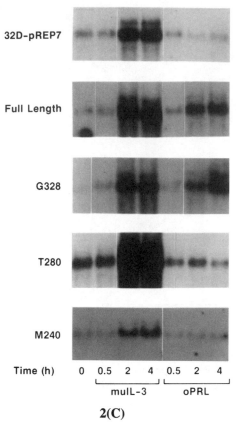

2(C)

Figure 2. Prolactin receptor mutagenesis establishes a requirement for membrane proximal regions of the PRL receptor in order to support PRL-dependent proliferation and gene transcription.

(A) Deletion mutants of the PRL receptor are indicated schematically and are represented along with the full-length receptor. (B) Proliferation response of PRL receptor mutants in response to IL-2 or human-PRL. Deletion of the Box 2 homology domain results in the loss of PRL-dependent proliferation. (C) Induction of ornithine decarboxylase (ODC) mRNA is a measure of the PRL receptor mutant's ability to transduce signal to the nucleus. ODC message was measured using an ODC specific ^{32}P-labelled probe on RNA from cells stimulated with 1 nM PRL or IL-3 for 0.5, 2, or 4 hr. 32D-pREP7 is vector alone, and the various PRL receptor mutants are designated on the left as per panel A. (Reproduced with permission from the *Journal of Biological Chemistry.*)

IL-2 receptor β and γ_c chains in order for these receptors to support JAK association, JAK kinase activation, and ligand-dependent signal transduction (Malabarba *et al.*, 1995; Miyazaki *et al.*, 1994; Russell *et al.*, 1994; Howard *et al.*, 1995).

These studies thus suggest that JAK family kinase activation resulting from cytokine-induced receptor aggregation is the event that initiates receptor level

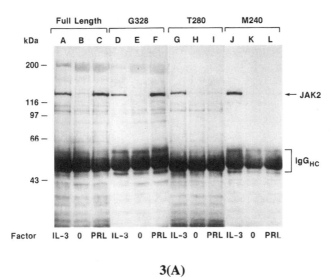

3(A)

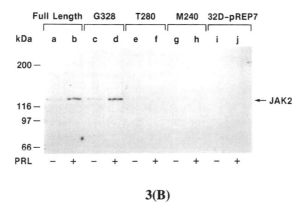

PRL-Induced Binding of JAK2 to the PRLR

3(B)

Figure 3. Loss of PRL receptor signaling ability correlates with a loss of JAK2 activation and receptor-association.

(A) Anti-phosphotyrosine immunoblot of JAK2-immunoprecipitated proteins from lysates of 32D clonal variants after hormone exposure. Cells were incubated with IL-3 (100 nM), medium (0), or PRL (100nM) for 5 min prior to analysis. (B) Anti-JAK immunoblot of PRL/PRL receptor complexes from lysates of 32D clones expressing various PRL receptors or vector alone (32D-pREP7). Cells were incubated with (+) or without (−) 10 nM biotinylated-PRL, and activated receptor complexes were purified by streptavidin-agarose beads and analyzed by JAK2 immunoblotting. (Reproduced with permission from the *Journal of Biological Chemistry*.)

signal transduction. Support for this can be found from studies in other cytokine receptor systems, which have also shown a requirement for membrane proximal domains of the receptor to mediate JAK activation, which directly correlates with cytokine-induced signal transduction (Quelle *et al.*, 1994; Witthuhn *et al.*, 1993; Narazaki *et al.*, 1994; VanderKuur *et al.*, 1994). Furthermore, work by Watling *et al.* (1993), which shows that JAK2 restores interferon γ signal transduction in an unresponsive cell line, clearly established the requirement for JAK activation as a signal transduction initiating mechanism.

While JAK activation appears to be obligatory for cytokine induced signal transduction, the manner by which these kinases are activated varies among receptors. Some receptor systems associate with and activate a single JAK family member. This is exemplified by JAK2 activation in response to PRL and GH (Rui *et al.*, 1994b; Argetsinger *et al.*, 1993). Other systems, such as those for IL-2, IL-4, IL-7, IL-9, interferon γ, CNTF, LIF, OSM, and IL-6, which contain multiple receptor chains, activate multiple JAKs (Miyazaki *et al.*, 1994; Russell *et al.*, 1994; Witthuhn *et al.*, 1994; Yin 95; Bacon *et al.*, 1995; Stahl *et al.*, 1994; Silvennoinen *et al.*, 1993; Musso *et al.*, 1995).

In these multiple subunit systems, there is a clear lack of symmetry regarding the extent to which different JAKs become activated. In the IL-2 receptor system, JAK1 associates with the β chain and JAK3 associates with the γ_c chain. While both JAK1 and JAK3 are activated by IL-2, JAK3 activity is induced to levels almost two- to three-fold higher than that of JAK1 (Evans, unpublished results). Similarly, in the IL-4 receptor system, which also signals through JAK1 and JAK3, it has been shown that JAK3 activation is the dominant event and that in some cell types JAK3 activation may be the only requirement for the initiation of signal transduction (Malabarba *et al.*, 1995).

Even though these systems demonstrate a form of specific JAK dominance in multiple JAK family kinase containing receptor complexes, a great deal of evidence establishes the absolute requirement for two or more JAKs in order to propagate signal in multi-chain receptor systems. The nature of this phenomenon stems from the preferential association of distinct JAK family members with specific receptor subunits, and implies a requirement for a functional interaction of JAK kinases within an aggregated receptor complex in order to mediate JAK activation.

3. How Does Receptor Aggregation Induce JAK Activation?

It has been proposed that cytokine induced homo- or heterodimerization of receptor subunits induces JAK association, resulting in JAK transphosphorylation and increased activity in a manner analogous to the activation of receptor tyrosine kinases (Stahl and Yancopoulos, 1993). Support for this can be seen upon analysis of recombinant rat JAK2 produced via the baculovirus expression system (rJAK2)

(Duhe and Farrar, 1995). Expression of rJAK2 in Sf21 cells produces a 120 kDa protein that is recognized by both anti-JAK2 and anti-phosphotyrosine antibodies. Low level expression of rJAK2 does not result in significant rJAK2 tyrosine phosphorylation. This can only be seen at higher levels of rJAK2 expression, suggesting that activation of rJAK2 occurs following an increased intracellular concentration of the enzyme due to overexpression (Figs. 5 and 6). This suggests that JAK2 activation in this system is the result of the expression of JAK2 reaching some threshold level at which JAK2/JAK2 interaction can lead to kinase activation. The common interpretation is that JAK2 transphosphorylation induces tyrosine kinase activity in a manner analogous to the activation of receptor tyrosine kinases. However, these data do not preclude the possibility that JAK2 structural domains may influence activity in a manner that may or may not involve transphosphorylation.

To address the functional importance of various domains of JAK2, several mutant forms of rJAK2 were generated by deletion mutagenesis and analyzed for enzyme activity and their ability to coordinately regulate JAK activity. A schematic representation of JAK2 is shown in Fig. 4. A characteristic of this family is the presence of two regions that are homologous to the tyrosine kinase catalytic domain: a pseudo-catalytic domain within the JH2 domain and the functional catalytic domain contained within the JH1 domain near the carboxy-terminus (Wilkes *et al.*, 1991). As depicted in Fig. 4, there are additional homology domains that are well conserved within the family.

Several JAK2 mutants were generated and designated (NΔ661)rJAK2, an amino-terminus deletion mutant lacking amino acids 1-661, and (CΔ795)rJAK2, a carboxy-terminus deletion beginning at amino acid 795 (Fig. 4). Expression of the (CΔ795)rJAK2 mutant in Sf21 insect cells did not result in detectable tyrosine phosphorylation of this protein as determined by JAK2 immune precipitation and anti-phosphotyrosine immunoblot, which verifies that removal of the carboxy-terminal catalytic domain abrogates JAK activity (Fig. 5). Interestingly, however, expression of (NΔ661)rJAK2, which contains the JH1 catalytic domain and part of the JH2 domain, resulted in the production of a hyperactive form of JAK2 (Fig. 6). Thus, at similar levels of expression, the (NΔ661)rJAK2 protein, lacking the amino-terminal domains, was phosphorylated to levels that appeared to be at least 10-fold higher than full-length rJAK2 (Fig. 6). Furthermore, this JAK2 mutant did not require an extended period of culture in order to elevate protein production and increase kinase activity, as was observed using full-length rJAK2 (Figs. 5 and 6). Hyperactivity of (NΔ661)rJAK2 is further supported with *in vitro* kinase assays, which show a dramatic increase in the autophosphorylating activity of this mutant when compared to full-length JAK2 results in increased phosphorylation both *in vivo* and *in vitro*, and strongly suggests a role for the amino-terminal domain in controlling JAK2 activity.

Further analysis suggests a distinct intermolecular role of JAK family kinases in regulating the activity of the JAK pool. Co-expression of the inactive mutant,

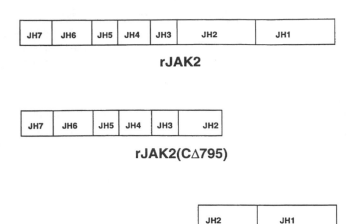

Figure 4. Schematic representation of JAK family kinases.

JAK family tyrosine kinases contain various conserved domains denoted here as JAK Homology domains (JH1-JH7). The bonafide catalytic domain is contained within JH1 and the pseudocatalytic domain within JH2. Analysis of amino acid similarities within this family revealed that the JH1 and JH2 domains have the highest level of conservancy (approx. 70–75%) among family members, with the JH3-JH7 domains possessing approx. 40–50% similarity. Deletion mutants of cloned rat JAK2 were generated (rJAK2(CΔ795) and (NΔ661)rJAK2), and are depicted schematically below full-length rJAK2. (Reproduced with permission from the *Journal of Biological Chemistry*.)

(CΔ795)rJAK2, with full-length rJAK2 results in a decrease in the *in vivo* phosphorylation of rJAK2 without detectable phosphorylation of the inactive form when compared to the expression of rJAK2 alone (Fig. 5). This suggests a dominant negative role for the amino-terminal domains of JAK2 in controlling JAK2 activity. The dominant role of the amino-terminus in regulating JAK2 activity is further supported when co-expression of the hyper-active mutant, (NΔ661)rJAK2, and full-length of rJAK2 does not result in enhanced tyrosine phosphorylation of the wild type enzyme (Fig. 6). These data strongly suggest the absence of intramolecular catalytic activity associated with these domains are involved in JAK2/JAK2 interaction, and that in the absence of intramolecular catalytic activity associated with these domains their function is dominantly negative. Further, a mutant form of *hopscotch*, a JAK family kinase from *Drosophilia*, was recently isolated which possesses a point mutation in the amino terminal portion of the enzyme (Harrison *et al.*, 1995; Luo *et al.*, 1995). This results in the production of a constitutively active kinase and further supports the contention that JAK activity may be regulated at one or more levels which may involve the intra- or intermolecular interaction of JAK2 with an inhibitory domain contained within the amino-terminus.

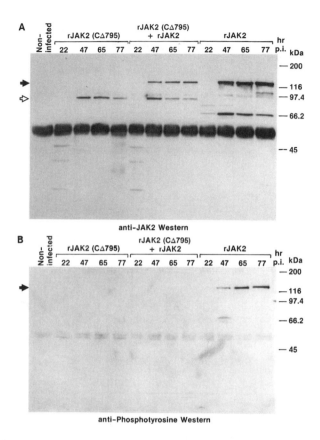

Figure 5. Analysis of production and tyrosine phosphorylation of full-length rJAK2 and rJAK2(CΔ795).

Insect cells (Sf21) were infected either individually or simultaneously with recombinant baculoviruses expression rJAK2 and rJAK2(CΔ795). A control plate was left uninfected. Samples were harvested 22 hr, 47 hr, 65 hr, or 77 hr post infection as indicated. Cells were lysed and immunoprecipitated with anti-JAK2. Immune precipitates were analyzed for JAK expression by anti-JAK2 immunoblot (panel A) and tyrosine phosphorylation by anti-phosphotyrosine immunoblot (panel B). The black arrows point to the 120 kDa rJAK2; the white arrows to the 90 kDa rJAK2(CΔ795). (Reproduced with permission from the *Journal of Biological Chemistry.*)

The role that JAK transphosphorylation plays in regulating JAK activity is not well understood. If transphosphorylation were a dominant mechanism regulating enzyme activity one would predict that co-expression of the hyper-active mutant (NΔ661)rJAK2 and rJAK2 would result in increased phosphorylation of the full-length rJAK2 molecule. This is clearly not the case (Fig. 6), and

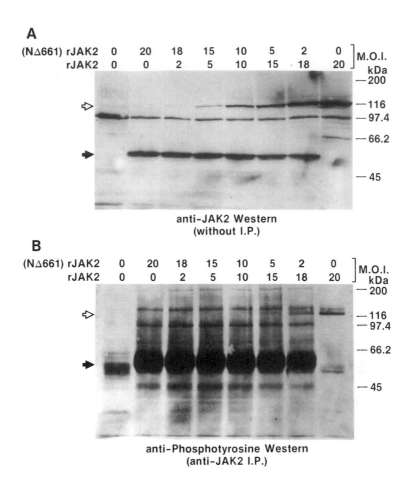

Figure 6. Analysis of production and tyrosine phosphorylation of full-length rJAK2 and (NΔ661)rJAK2.

Sf21 insect cells were infected with recombinant baculovirus containing full-length rJAK2 or (NΔ661)rJAK2 and analyzed for expression (panel A) and tyrosine phosphorylation (panel B), as described in Fig. 5. (Reproduced with permission from the *Journal of Biological Chemistry*.)

further strengths the argument that the regulation of JAK activity is primarily controlled by amino-terminus interactions. However, this does not preclude a role for transphosphorylation in regulating amino-terminus interactions.

The exact function of specific amino-terminus domains of JAK2 remains to be determined. In addition to the potential catalytic regulatory role of amino-terminal domains, Zhao *et al.* (1995) show that JAK2 interacts with the GM-CSF receptor β_c chain via amino acids 1-294 contained in the amino-terminus of JAK2. Using a carboxy-truncated form of JAK2, Zhuang *et al.* (1994) have inhibited EPO-dependent mitogenesis and JAK2 activation. This further supports a regulatory role for JAK amino-terminal domains in controlling receptor association and/or JAK activation. Together these data imply a multi-functional role of the JAK amino-terminal domains in regulating JAK activity and the ability to couple to cytokine receptor activation. While the results imply a regulatory role for these domains, the specific elucidation of domain function requires more detailed analysis.

4. Multiple Pathway Integration and Receptor Structural Motifs Define the Mitogenic Response

The role that receptor complex tyrosine phosphorylation plays in regulating mitogenic signal transduction within the cytokine receptor family is not well understood. Using the IL-2 receptor system as a model, several approaches were used to investigate the specific function of receptor cytoplasmic domains and tyrosine phosphorylation in regulating several key mitogenic pathways. These signal transduction pathways have been associated with the positive regulation of cytokine induced mitogenesis in one or more receptor systems and include the JAK/STAT (reviewed by Ihle *et al.*, 1994), ras activation (reviewed by Maruta and Burgess, 1994; Burgering and Bos, 1995), P13 kinase (reviewed by Kapeller and Cantley, 1994), and raf/map kinase (reviewed by Daum *et al.*, 1994; Avruch *et al.*, 1994) pathways.

It is well established that a consequence of cytokine induced activation of JAK family kinases is tyrosine phosphorylation of cytokine receptor chains. The biological significance of receptor tyrosine phosphorylation is well documented within the receptor tyrosine kinases. Within these receptors, tyrosine phosphorylation of the receptor functions to couple tyrosine kinase activity to downstream pathway activation. This is directed by the specificity by which distinct pathway-coupling proteins recognize and bind to tyrosine-phosphorylated motifs within the cytoplasmic domain of the receptor (Stahl *et al.*, 1995). It thus becomes important to establish the role that cytoplasmic domain structural motifs, especially those containing phosphorylated tyrosine residues, play in regulating downstream signal transduction within the cytokine receptor superfamily.

As previously mentioned, membrane proximal domains of cytokine receptors have been shown to function in JAK family kinase association, and this association is required for cytokine-dependent signal transduction. To further investigate this, JAK/receptor function in the IL-2 receptor system was investigated in HT-2 cells possessing various chimeric forms of EPO/IL-2 receptor combinations. This system has previously proved useful in establishing the importance of receptor extracellular domains for ligand binding and receptor aggregation and the function of receptor cytoplasmic domains in directing aggregation induced signal transduction

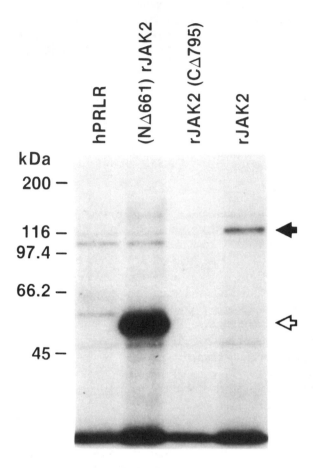

Figure 7. In vitro kinase assay of various JAK2 deletion mutants.

Sf21 cells were infected with recombinant baculovirus expression rJAK2, rJAK2(CΔ795), or (NΔ661)rJAK2. Cells were harvested 112 hr post infection, lysed, and immune precipitated with anti-JAK2 antiserum. Immune precipitates were washed and subjected to *in vitro* kinase assays. Autophosphorylation was determined by gel electrophoresis and autoradiography. (Reproduced with permission from the *Journal of Biological Chemistry*.)

(Goldsmith *et al.*, 1995; Gaffen *et al.*, 1995). Co-transfections of EPOβ and EPOγ_c are clearly capable of supporting EPO-dependent mitogenesis as well as EPO-dependent activation of JAK1 and JAK3 (Lai *et al.*, 1995). In similar experiments, HT-2 cells were co-transfected with EPOβ and an EPO receptor mutant containing an intact extracellular domain but with a severely truncated cytoplasmic domain containing the membrane proximal JAK2 binding domain but lacking the carboxy-terminal 162 amino acids involved in STAT and SH2 domain interactions (Lai *et al.*, 1995). EPO treatment of these cells resulted in induction of cell growth and activation of JAK1 and JAK2, indicating that this receptor combination retained the ability to transduce signal (Lai *et al.*, 1995). Expression of a more severely truncated form of the EPO receptor lacking the JAK2 binding domain, or expression of a mutant that cannot activate JAK2 kinase function, failed to reconstitute signal in EPOβ transfected HT-2 cells (Lai *et al.*, 1995). This supports the contention that receptor cytoplasmic domain structure dictates JAK association, and suggests that the role of the γ_c chain may simply be to supply a functional JAK to the receptor heterodimer, allowing JAK interaction and tyrosine kinase activation.

It has previously been shown that the γ_c chain is rapidly tyrosine phosphorylated following cytokine treatment (Asao *et al.*, 1992). To further test the importance of γ_c chain tyrosine phosphorylation in the regulation of mitogenic signal, and EPOγ_c chain chimera (EPOγYF) containing point mutations of all cytoplasmic tyrosines to phenylalanine was constructed (Goldsmith *et al.*, 1995). HT-2 cells co-transfected with EPOβ and EPOγYF were able to fully support EPO-induced mitogenesis, suggesting a lack of involvement of γ_c chain tyrosine phosphorylation in directing cytokine induced cell growth.

Receptor tyrosine phosphorylation has been shown to be critical in determining the pathways that are capable of coupling to growth factor stimulation. The observation that γ_c chain tyrosine phosphorylation is not required to support mitogenic signal suggests that in the IL-2 receptor model, the IL-2 receptor β chain plays a dominant role in regulating downstream pathway coupling. To further investigate this, deletion mutants of the IL-2 receptor β chain were generated, transfected into Ba/F3 cells, and analyzed for their ability to support mitogenesis and signal pathway coupling (Fig. 8). This analysis indicates that FL, AD, and BD mutants were capable of supporting JAK activation, tyrosine phosphorylation of multiple proteins, and mitogenesis (Howard *et al.*, 1995). The ability to couple to signaling pathways associated with tyrosine kinase activation was thus investigated using these receptor mutants that function to activate JAK kinases.

Cytokine-dependent tyrosine phosphorylation of a family of proteins known as STAT (Signal Transducers and Activators of Transcription) proteins has recently defined a specific mechanism by which signal can be propagated from the activated receptor complex directly to the nucleus. STAT receptor association and tyrosine phosphorylation generates an activated STAT complex that will interact specifically with several DNA sequences that were originally identified in interferon signaling studies. These have thus been referred to as GAS (γ-interferon activation sequences) or GRR (γ-interferon response regions). Nucleotide probes of these sequences have been used in a gel mobility shift assay to identify whether cytokine stimulation induces the formation of STAT DNA-binding activity. In the IL-2 receptor system, a GRR sequence and gel mobility shift assay was used to identify the activation of STAT factor DNA binding activity in the receptor mutants

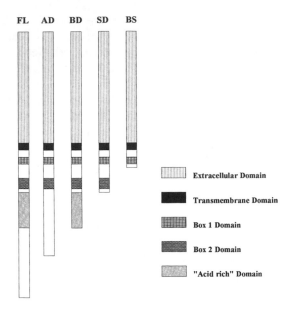

8(A)

IL-2-dependent Proliferation of IL-2R beta mutants

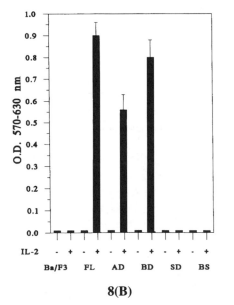

8(B)

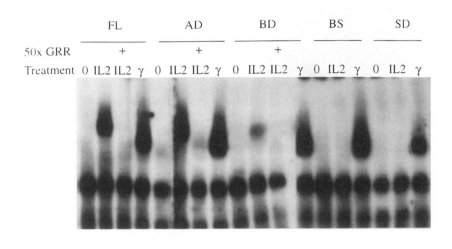

8(C)

Figure 8. IL-2 receptor β chain deletion-mutagenesis defines regions critical for mitogenesis and coupling to STAT DNA-binding activity.

(A) Deletion mutants of the IL-2 receptor β chain were generated and are represented schematically in panel A along with the position of cytoplasmic tyrosine residues. (B) Receptors were transfected into murine BaF3 cells and assayed for proliferative activity in response to 1 nM IL-2 or IL-3. (C) Whole cell extracts from IL-2 receptor β chain mutant cell lines were assayed for complex formation with the GRR probe by electrophorectic mobility shift assay. Cells were treated with 1 ug/ml IL-2 (IL-2), 10 ng/ml interferon γ (γ), or basal medium (0). Specificity of complex formation was demonstrated by competition with 50-fold excess GRR. (Reproduced with permission from the *Biochemical Journal.*)

that induce tyrosine phosphorylation (FL, AD, and BD mutants) (Howard *et al.*, 1995). This analysis reveals that FL and AD receptors support GRR-complex formation equally well (Fig. 8C). BD receptors, however, showed a much reduced induction of nuclear complex formation in response to IL-2 (Fig. 8C). Because STAT phosphorylation requires initial association with a tyrosine phosphorylated receptor, these results indicate that IL-2-dependent STAT activation may involve phosphorylation of tyrosine 392 and/or tyrosine 510, which are deleted in the BD mutants (Fig. 8C). Gaffen *et al.* (1995) and Lin *et al.* (1995) support this contention by demonstrating that tyrosine 392 and/or 510 of the IL-2 receptor β chain mediate STAT-5 binding and activation. However, the reduction, but not elimination, of GRR nuclear complex formation in the BD mutants that lack these tyrosines suggests that more proximal tyrosines may also be involved in mediating this effect.

Inference can be drawn concerning the role that STAT activation plays in regulating mitogenesis. These results suggest that STAT activation is not critical in inducing mitogenesis, since the BD mutant can support cell growth to levels identical to the wild type receptor while showing a severe reduction in STAT complex formation (Howard *et al.*, 1995). Furthermore, studies in the EPO and IL-3 receptor systems have shown that mutants that lack the STAT binding domain are still capable of responding mitogenically to cytokine, suggesting that STAT activation does not play a major role in regulating the mitogenic response.

In contrast to STAT pathway signaling, the activation of Ras and downstream targets has been shown to be fundamentally important to growth factor induced cell growth. Within the cytokine receptors, however, Ras has been shown to play a variable role in mitogenic signaling. EPO, IL-3, and IL-2 receptor mutants that have lost the ability to activate Ras still function to support mitogenesis. However, mutants of the GM-CSF receptor, which do not activate Ras have been shown to require a basal level of Ras activity supplied by serum in order to support proliferation following GM-CSF treatment (Sakamaki and Yonehara, 1994). This suggests that Ras pathway activation may indeed be important in regulating the maximal proliferative response to cytokine.

It has been well established that receptor level activation of Ras involves the induction of adaptor pathways that mediate Ras activation by controlling GNRF activity (reviewed by Boguski and McCormick, 1993). Because the induction of adaptor pathway function may also allow the activation of distinct Ras-like pathways, it is important to determine the cytokine receptor mechanisms involved in regulating adaptor pathway activation. As Shc receptor binding and tyrosine phosphorylation has been shown to initiate the Shc/Grb2/Sos adaptor pathways, IL-2 receptor mutants were assayed for their ability to support Shc phosphorylation and receptor association. Previous analysis has shown that only FL, AD, and BD receptors are capable of supporting JAK kinase activation and substrate phosphorylation in response to IL-2 (Howard *et al.*, 1995). However, even though AD receptors support IL-2-induced tyrosine phosphorylation they failed to induce Shc tyrosine phosphorylation (Fig. 9A). Furthermore, analysis of receptor association indicates that Shc binding requires a motif (amino acid 315-384) which is deleted in the AD mutant (Fig. 9B). Using IL-2 receptor mutants containing tyrosine to phenylalanine mutations revealed that a motif surrounding phosphorylated tyrosine 338 mediates Shc binding and tyrosine phosphorylation (Fig. 10). This supports previous results indicating that this region may be important in mediating Ras pathway induction in response to IL-2 (Satoh *et al.*, 1992). Because AD mutants support IL-2-dependent mitogenesis, these results suggest that adaptor pathway activation is not a prerequisite for cell growth. AD receptors have also been shown to support mitogenesis in serum-free and insulin-free conditions (Evans *et al.*, 1995), which further supports the lack of involvement of these pathways in the control of proliferation.

This work and results from other cytokine receptor systems have established that the elimination of either adaptor pathway activation, Ras activation, or STAT activation are not universally associated with a reduction in cytokine-dependent

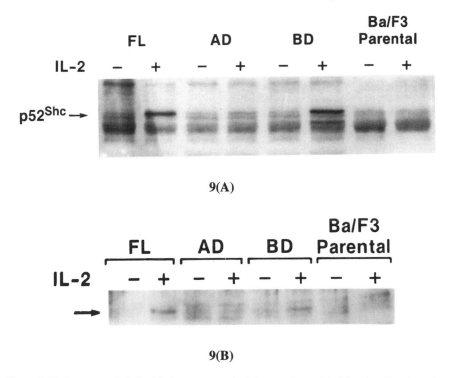

9(A)

9(B)

Figure 9. IL-2 receptor β chain deletion mutagenesis defines regions critical for signaling through SHC phosphorylation and receptor association.

A) Ba/F3 cells containing FL, AD, or BD receptors were treated with or without IL-2 (+ or −) for 15 min, lysed, and immune precipitated with anti-Shc antibody. Relative tyrosine phosphorylation of Shc was determined by anti-phosphotyrosine immunoblot of anti-Shc immune precipitates. The position of p52^{Shc}, is indicated on the left. B) Ba/F3 cells containing FL, AD, or BD receptors were treated and lysed as in panel A, and immune precipitated with anti-IL-2Rβ chain antibody conjugated to protein-A Sepharose. Immune precipitates were probed for the presence of Shc by anti-Shc immunoblot. The position of p52^{Shc}, is indicated on the left. Note that Shc interacts with the acid rich domain that is deleted in the AD mutants and this association is required in order to mediate IL-2-dependent Shc phosphorylation. (Reproduced with permission from the *Journal of Biological Chemistry.*)

proliferation What then are the critical signal transduction pathways utilized by cytokine to induce mitogenesis? Several groups have shown that the elimination of Raf activation results in a universal block to proliferation induced by cytokine (Muszynski *et al.*, 1995; Brennschedit *et al.*, 1994; Tornkvist *et al.*, 1994; Riedel *et al.*, 1993). This suggests that Raf activation and the modulation of downstream pathways, such as the MAP kinase cascade, are critical in regulating mitogenesis.

A common mechanism by which many mitogenic factors modulate Raf function is via Ras pathway activation. Because it has been shown that Ras pathway activation is not essential for cell growth by cytokine, it is plausible that these receptor systems may use alternate mechanisms that impact on the Raf/MAP kinase pathways. Raf has been shown to directly associate with the IL-2 receptor β chain (Maslinski et al., 1993), to be activated in response to IL-2 (Turner et al., 1991, 1993) and to be activated by direct tyrosine phosphorylation in other systems (Fabian et al., 1993). This indicates that Raf function may be directly modulated at the level of cytokine receptor binding and tyrosine phosphorylation and may not require Ras function to mediate Raf mitogenic signal. Upon analysis of IL-2 receptor deletion mutants, Raf was found to interact with the FL and BD mutant; however, RAf interaction was greatly diminished in the AD mutant (Evans, unpublished results). If the Raf/MAP kinase pathway is indeed critical in regulating cell growth then this indicates the possible existence of additional Raf coupling mechanisms.

A recent report by Karnitz et al. (1995) indicates that IL-2 activation of PI3 kinase mediates the downstream activation of MEK and MAP kinases in a manner that does not require Raf function. PI3 kinase activation has further been shown to require tyrosine phosphorylation of the p85 subunit (Karnitz et al., 1994), and this protein has been found associated with the IL-2 receptor β chain (Truitt et al., 1994). Structure-function analysis of PI3 kinase association with the IL-2 receptor β chain has shown that this protein binds phosphorylated tyrosine 392 via an SH2 domain located in the p85 subunit (Truitt et al., 1994). AD mutants retain the cytoplasmic domain containing tyrosine 392, and thus may functionally couple to PI3 kinase activation. This could explain why AD mutants are capable of supporting mitogenesis in the absence of additional mechanisms that could couple to mitogenic pathways.

Taken together, these studies of mitogenic-pathway coupling in the IL-2 receptor system describe a complex system with the potential for utilizing several redundant pathways for the induction of mitogenesis (Fig. 11). It is clear that the ability of cytokine to couple to these pathways is dependent on receptor aggregation, JAK activation, and receptor tyrosine phosphorylation. However, these studies fail to identify the single critical pathway emanating from the receptor. The full-length IL-2 receptor β chain (FL) is potentially capable of inducing mitogenesis through direct receptor-level activation of Raf, adaptor/Ras pathway activation, or phosphorylation and activation of PI3 kinase. Because the AD mutant retains the PI3 kinase binding motif (tyrosine 392), does not activate Ras, and has reduced Raf binding, the PI3 kinase/MEK pathway appears as the dominant mechanism. BD mutants activate Ras and Raf and thus easily establish mitogenic pathway induction. Additional structure-function analysis of the IL-2 receptor β chain supports redundant pathway activation by cytokine in the induction of growth signal (Goldsmith et al., 1994).

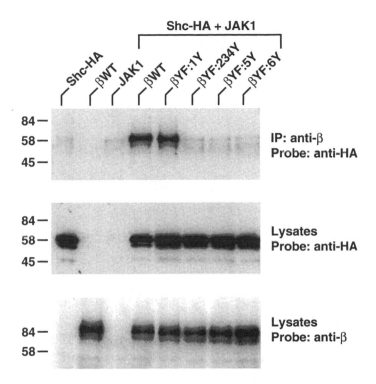

Figure 10. IL-2 receptor β chain point mutations of tyrosine to phenylalanine establish the motif surrounding phosphorylated tyrosine 338 as the Shc binding site within the IL-2 receptor β chain.

COS-7 cells were transfected with constructs containing Shc-HA, Jak-1 and/or wild type IL-2Rβ chain (βWT), or tyrosine to phenylalanine mutants of the IL-2Rβ chain containing only tyrosine 338 (βYF:1Y), tyrosines 355, 358, and 361 as a group (βYF:234Y), tyrosine 392 (βYF:5Y), or tyrosine 510 (βYF:6Y). Cells were cultured, lysed, and immune precipitated with antibody specific for the IL-2 receptor β chain and probed for Shc-HA association using anti-HA immunoblot (upper panel). To verify equivalent expression in transfected cells, lysates were subjected to anti-HA immunoblot (middle panel) or anti-β chain immunoblot (lower panel). (Reproduced with permission from the *Journal of Biological Chemistry.*)

It is clear that the mechanism utilized by cytokine to promote cell growth capitalizes on the availability of several mitogenic pathways. Upon activation, any one of these can result in the propagation of some degree of growth stimulus. It is likely that the normal physiological response of a given cell type to cytokine is ultimately governed by the quality of these pathways and their ability to be induced. This model of redundant pathway utilization is therefore strictly dependent on the cellular phenotype, and explains the varied mitogenic response of different cell types to the same cytokine.

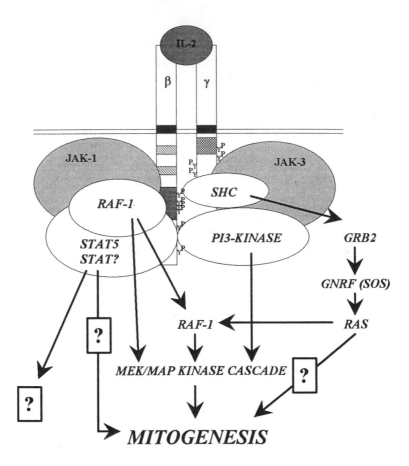

Figure 11. The model of IL-2 signal transduction describes the use of redundant pathway activation for the induction of mitogenesis.

Mitogenic signal transduction induced by IL-2 capitalizes on the availability of redundant pathways which can couple to growth promotion. Signal pathway activation is dependent on the tyrosine phosphorylation of the IL-2 receptor β chain and activation of receptor associated proteins.

References

Argetsinger, L.S., Campbell, G.S., Yang, X., Witthuhn, B.A. Silvennoinen, O., Ihle, J.N., and Carter-Su, C., 1993, Identification of JAK2 as a growth hormone receptor-associated tyrosine kinase, *Cell*, **74**: 237–244.

Asao, H., Kumaki, S., Takeshita, T., Nakamura, M., and Sugamura, K., 1992, IL-2-dependent *in vivo* and *in vitro* tyrosine phosphorylation of IL-2 receptor γ chain, *FEBS Lett*, **304**:141–145.

Avruch, J., Zhang, X.F., and Kyriakis, J.M., 1994, Raf meets Ras: completing the framework of a signal transduction pathway, *Trends Biochem. Sci.*, **19**:279–283.

Bacon, C.M., McVicar, D.W., Ortaldo, J.R., Rees, R.C., O'Shea, J.J., and Johnston, J.A., 1995, Interleukin 12 (IL-2) induces tyrosine phosphorylation of JAK2 and TYK2: Differential use of Janus family tyrosine kinases by IL-2 and IL-12, *J. Exp. Med.*, **181**:399–404.

Bazan, J.F., 1990, Structural design and molecular evolution of a cytokine receptor superfamily, *Proc. Natl. Acad. Sci. USA*, **87**:6934–6938.

Bogusky, M.S., and McCormick, F., 1993, Proteins regulating Ras and its relatives, *Nature*, **366**:643–653.

Brennscheidt, U., Riedel, D., Kolch, W., Bonifer, R., Brach, M.A., Ahlers, A., Mertelsmann, R.H., and Herrmann, F., 1994, Raf-1 is a necessary component of the mitogenic response of the human megakaryoblastic leukemia cell line MO7 to human stem cell factor, granulocyte-macrophage colony-stimulating factor, interleukin 3 and interleukin 9, *Cell Growth Differ*, **5**:367–372.

Burgering, B.M., and Bos, J.L., 1995, Regulation of Ras-mediated signalling: more than one way to skin a cat, *Trends Biochem. Sci.*, **20**:18–22.

Cosman, D., 1993, The hematopoietin receptor superfamily, *Cytokine*, **5**:95–106.

Cunningham, B.C., Ultsch, M., de Vos, A.M., Mulkerrin, M.G., Clausner, K.R., and Wells, J.A., 1991, Dimerization of the extracellular domain of the human growth hormone receptor by a single hormone molecule, *Science*, **254**:821–825.

DaSilva, L., Howard, O.M.Z., Rui, H., Kirken, R.A., and Farrar, W.L., 1994, Growth signaling and JAK2 association mediated by membrane-proximal cytoplasmic regions of prolactin receptors, *J. Biol. Chem.*, **269**:18267–18270.

Daum, G., Eisenmann-Tappe, I., Fries, H.W., Troopmair, J., and Rapp, U.R., 1994, The ins and outs of Raf kinase, *Trends Biochem. Sci.*, **19**:474–480.

de Vos, A.M., Ultsch, M., and Kossiakoff, A.A., 1992, Human growth hormone and extracellular domain of its receptor: Crystal structure of the complex, *Science*, **255**:306–312.

Duhe, R.J., and Farrar, W.L., 1995, Characterization of active and inactive forms of the JAK2 protein-tyrosine kinase produced via the baculovirus expression vector system, *J. Biol. Chem.*, (in press).

Evans, G.A., Goldsmith, M.A. Johnston, J.A., Xu, W., Weiler, S.R., Erwin, R., Howard, O.M.Z., Abraham, R.T., O'Shea, J.J., Greene, W.C., and Farrar, W.L., 1995, Analysis of IL-2-dependent signal transduction through the SHC/GRB2 adaptor pathway: IL-2-dependent mitogenesis does not require Shc phosphorylation or receptor association, *J. Biol. Chem.*, (in press).

Fabian, J.R., Daar, I.O., and Morrison, D.K., 1993, Critical residues regulate the enzymatic and biological activity of Raf-1 kinase, *Mol. Cell. Biol.*, **13**:7170–7179.

Gaffen, S.L., Lai, S.Y., Xu, W., Gouilleux, F., Groner, B., Goldsmith, M.A., and Greene, W.C., 1995, Signaling through the interleukin 2 receptor β chain activates a STAT-5-like DNA-binding activity, *Proc. Natl. Acad. Sci.*, **92**:7192–7196.

Giri, J.G., Anderson, D.M., Kumaki, S., Park, L.S., Grabstein, K.H., and Cosman, D., 1995, IL-15, a novel T cell growth factor that shares activities and receptor components with IL-2, *J. Leukoc. Biol.*, **57**:763–766.

Goldsmith, M.A., Xu, W., Amaral, M.C., Kuczek, E.S., and Greene, W.C., 1994, *J. Biol. Chem.*, **269**:14698–14704.

Goldsmith, M.A., Lai, S.Y., Xu, W., Amaral, C., Kuczek, E.S., Parent, L.J., Mills, G.B., Tarr, K.L., Longmore, G.D., and Greene, W.C., 1995, Growth signal transduction by the human interleukin-2 receptor requires cytoplasmic tyrosines of the β chain and non-tyrosine residues of the γ_c chain, *J. Biol. Chem.*, **270**:27129–21737.

Harrison, D.A., Binari, R., Nahreini, T.S., Gilman, M., and Perrimon, N., 1995, Activation of a *Drosophila*, Janus kinase (JAK) causes hematopoietic neoplasia and developmental defects, *EMBO J.*, **14**:2857–2865.

Heldin, C., 1995, Dimerization of cell surface receptors in signal transduction, *Cell*, **80**:213–223.

Howard, O.M.Z., Kirken, R.A., Garcia, G.G., Hackett, R.H., and Farrar, W.L., 1995, Structural domains of the interleukin-2 receptor β critical for signal transduction: kinase association and nuclear complex formation, *Biochem. J.*, **306**:217–224.

Ihle, J.N., Witthuhn, B.A., Quelle, F.W., Yamamoto, K., Thierfelder, W.E., Kreider, B., and Silvennoinen, O., 1994, Signaling by the cytokine receptor superfamily: JAKs and STATs, *Trends Biochem. Sci.*, **19**:222–227.

Johnston, J.A., Kawamura, M., Kirken, R.A., Chen, Y.Q., Blake, T.B., Shibuya, K., Ortaldo, J.R., McVicar, D.W., and O'Shea, J.J., 1994, Phosphorylation and activation of the Jak-3 Janus kinase in response to interleukin-2, *Nature*, **370**:151–153.

Kapeller, R., and Cantley, L.C., 1994, Phosphatidylinositol 3-kinase, *Bioessays*, **16**:565–576.

Karnitz, L.M., Sutor, S.L., and Abraham, R.T., 1994, The Src-family kinase, Fyn, regulates the activation of phosphatidylinositol 3-kinase in an interleukin 2-responsive T cell line, *J. Exp. Med.*, **179**:1799–1808.

Karnitz, L.M., Burns, L.A., Sutor, S.L., Blenis, J., and Abraham, R.T., 1995, Interleukin-2 triggers a novel phosphatidylinositol 3-kinase-dependent MEK activation pathway, *Mol. Cell. Biol.*, **15**:3049–3057.

Kelly, P.A., Djiane, J., Postal-Vinay, M.C., and Edery, M., 1991, The proclactin/growth hormone receptor family, *Endocr. Rev.*, **12**:235–251.

Kimura, Y., Takeshita, T., Kondon, M., Ishii, N., Nakamura, M., Van-Shick, J., and Sugamura, K., 1995, Sharing of the IL-2 receptor gamma chain with the functional IL-9 receptor complex, *Int. Immunol.*, **7**:115–120.

Kirken, R.A., Rui, H., Howard, O.M.Z., and Farrar, W.L., 1994, Involvement of Jak-family tyrosine kinases in hematopoietin receptor signal transduction, *Progr. Growth Factor Res*, **5**:195–211.

Kondo, M., Takeshita, T., Ishii, N., Nakamura, M., Watanabe, S., Arai, K., and Sugamura, K., 1993, Sharing of the interleukin-2 (IL-2) receptor gamma chain between receptors for IL-2 and IL-4, *Science*, **262**:1874–1877.

Lai, S.Y., Xu, W., Gaffen, S.L., Lui, K.D., Longmore, G.D., Greene, W.C., and Goldsmith, M.A., 1995, The molecular role of the common γ_c subunit in signal transduction reveals functional asymmetry within multimeric cytokine receptor complexes, *Proc. Natl. Acad. Sci. USA*, (in press).

Lin, J.X., Migone, T.S., Tsang, M., Friedmann, M., Weatherbee, J.A., Zhou, L., Yamauchi, A., Bloom, E.T., Mietz, J., John, S., and Leonard, W.J., 1995, The role of shared receptor motifs and common STAT proteins in the generation of cytokine pleiotropy and redundancy by IL-2, IL-4, IL-7, IL-13, and IL-15, *Cell*, **2**:331–339.

Luo, H., Hanratty, W.P., and Dearolf, C.R., 1995. An amino acid substitution in the *Drosophila*, hop Tum-1 Jak kinase causes leukemia-like hematopoietic defects. *EMBO J.*, **14**:1412–1420.

Malabarba, M.G., Kirken, R.A., Rui, H., Koettnitz, K., Kawamura, M., O'Shea, J.J., Kalthoff, F.S., and Farrar, W.L., 1995, Activation of JAK3, but not JAK1, is critical to interleukin-4 (IL-4) stimulated proliferation and requires a membrane-proximal region of IL-4 receptor α, *J. Biol. Chem.*, **270**:8630–9637.

Maruta, H., and Burgess, A.W., 1994, Regulation of the Ras signalling network, *Bioessays*, **16**:489–496.

Maslinksi, W., Remillard, B., Tsudo, M., and Strom, T.B., 1993, Interleukin-2 receptor signal transduction: translocation of active serine-threonine kinase Raf-1 from IL-2 receptor into cytosol depends on IL-2 induced tyrosine kinase activation, *Transplant Proc.*, **25**:109–110.

Miyazaki, T., Kawahara, A., Fuji, H., Nakagawa, Y., Minami, Y., Liu, Z.J., Oishi, I., Silvennoinen, O., Witthuhn, B.A., Ihle, J.N., and Taniguchi, T., 1994, Functional activation of Jak1 and Jak3 by selective association with IL-2 receptor subunits, *Science*, **266**:1045–1047.

Mott, H.R., and Campbell, I.D., 1995, Four-helix bundle growth factors and their receptors: Protein-protein interactions, *Curr. Opin. Struct. Biol.*, **5**:114–121.

Musso, T., Johnston, J.A., Linnekin, D., Varesio, L., Rowe, T.K., O'Shea, J.J., and McVicar, D.W., 1995, Regulation of JAK3 expression in human monocytes: phosphorylation in response to interleukins 2, 4, and 7, *J. Exp. Med.*, **181**:1425–1431.

Muszynski, K.W., Ruscetti, F.W., Heidecker, G., Rapp, U., Troopmair, J., Gooya, J.M., and Keller, J.R., 1995, Raf-1 protein is required for growth-factor induced proliferation of hematpoietic cells, *J. Exp. Med.*, **181**:2189–2199.

Narazaki, M., Witthuhn, B.A., Yoshida, K., Silvennoinen, O., Yasukawa, K., Ihle, J.N., Kishimoto, T. and Tagar, T., 1994, Activation of JAK2 kinase mediated by the interleukin 6 signal transducer gp130, *Proc. Natl. Acad. Sci. USA*, **91**:2285–2289.

Nelson, B.H., Lord, D.J., and Greenberg, P.D. 1994, Cytoplasmic domains of the interleukin-2 receptor β and γ chainsmediate the signal for T-cell proliferation, *Nature*, **369**:333–336.

Quelle, F.W., Sato, N., Witthuhn, B.A., Inhorn, R.C., Eder, M., Miyajima, A., Griffin, J.D., and Ihle, J.N., 1994, JAK2 associates with the β$_c$ chain of the receptor for granulocyte-macrophage colony-stimulating factor, and its activation requires the membrane proximal region, *Mol. Cell. Biol.*, **14**:4335–4341.

Riedel, D., Brennscheidt, U., Kiehntopf, M., Brach, M., and Hermann, F., 1993, The mitogenic response of T cells to interleukin-2 requires Raf-1, *Eur. J. Immunol.*, **23**:3146–3150.

Rui, H., Lebrun, J.J., Kirken, R.A., Kelly, P.A., and Farrar, W.L., 1994a, JAK2 activation and cell proliferation induced by antibody mediated prolactin receptor dimerization, *Endocrinology*, **135**:1299–1306.

Rui, H., Kirken, R.A., and Farrar, W.L., 1994b, Activation of recpetor-associated tyrosine kinase JAK2 by prolactin, *J. Biol. Chem.*, **269**:5364–5368.

Russell, S.M., Johnston, J.A., Noguchi, M., Kawamura, M., Bacon, C.M., Friedmann, M., Berg, M., McVicar, D.W., Witthuhn, B.A., Silvennoinen, O., Goldman, A.S., Schmalstieg, F.C., Ihle, J.N., O'Shea, J.J., and Leonard, W.J., 1994, Interaction of the IL-2Rβ and γ$_c$ chains with Jak1 and Jak3: Implications for XSCID and XCID, *Science*, **266**:1042–1045.

Sakamaki, K., and Yonehara, S., 1994, Serum alleviates the requirement of the granulocyte-macrophage colony-stimulating factor (GM-CSF)-induced Ras activation for proliferation of BaF3 cells, *FEBS Lett.*, **353**:133–137.

Satoh, T., Minami, Y., Kono, T., Yamada, K., Kawahara, A., Taniguchi, T., and Kaziro, Y., 1992, Interleukin 2-induced activation of Ras requires two domains of interleukin-2 receptor β subunit, the essential region for growth stimulation and Lck-binding domain, *J. Biol. Chem.*, **267**:25423–25427.

Silvernnoinen, O., Ihle, J.N., Schlessinger, J., and Levy, D.E., 1993, Interferon-induced nuclear signalling by Jak protein tyrosine kinases, *Nature*, **366**:583–585.

Stahl, N., and Yancopoulos, G.D., 1993, The α's, β's and kinases of cytokine receptor complexes, *Cell*, **74**:587–590.

Stahl, N., Boulton, T.G., Farruggella, T., Ip, N.Y., Davis, S., Witthuhn, B.A., Quelle, F. W., Silvennoinen, O., Barbieri, G., Pellegrini, S., Ihle, J.N., and Yancopoulos, G.D., 1994, Association and activtion of Jak-Tyk kinases by CNTF-LIF-OSM-IL6 β receptor components, *Science*, **263**:92–95.

Stahl, N., Farruggella, T.J., Boulton, T.G., Zhong, Z., Darnell, J.E., and Yancopoulos, G.D., 1995, Choice of STATs and other substrates specified by modular tyrosine-based motifs in cytokine receptors, *Science*, **267**:1349–1353.

Tornkvist, A., Parpal, S., Gustavsson, J., and Stralfors, P., 1994, Inhibition of Raf-1 kinase expression abolishes insulin stimulation of DNA synthesis in H4IIE hepatoma cells, *J. Biol. Chem.*, **269**:13919–13921.

Truitt, K.E., Mills, G.B., Turck, C.W., and Imboden, J.B., 1994, SH2-dependent association of phosphatidylinositol 3′-kinase 85-kDa regulatory subunit with the interleukin-2 receptor β chain, *J. Biol. Chem.*, **269**:5937–5943.

Turner, B., Rapp, U., App, H., Greene, M., Dobashi, K., and Reed, J., 1991, Interleukin 2 induces tyrosine phosphorylation and activation of p72–74 Raf-1 kinase in a T cell line, *Proc. Natl. Acad. Sci. USA*, **88**:1227–1231.

Turner, B., Tonks, N.K., Rapp, U.R., and Reed, J.C., 1993, Interleukin 2 regulates Raf-1 kinase activity through a tyrosine phosphorylation-dependent mechanism in a T cell line, *Proc. Natl. Acad. Sci. USA*, **90**:5544–5548.

Ultsch, M., de Vos, A.M., and Kossiakoff, A.A., 1991, Crystals of the complex between human growth hormone and the extracellular domain of its receptor, *J. Mol. Biol.*, **22**:865–868.

VanderKuur, J.A., Wang, X., Zhang, L., Campbell, G.S., Allevato, G., Billestrup, N., Norstedt, G., and Carter-Su, C., 1994, Domains of the growth hormone receptor required for association and activation of JAK2 tyrosine kinase, *J. Biol. Chem.*, **269**:21709–21717.

Vigon, I., Mornon, J.P., Cocault, L., Mitjavila, M.T., Tambourin, P., Gisselbrecht, S., and Souyri, M., 1992, Molecular cloning and characterization of MPL, the human homolog of the v-mpl oncogene: Identification of a member of the hematopoietin growth factor receptor superfamily, *Proc. Natl. Acad. Sci. USA*, **89**:5640–5644.

Vita, N., Lefort, S., Laurent, P., Caput, D., and Ferrar, P., 1955, Characterization and comparison of the interleukin 13 receptor with the interleukin 4 receptor on several cell types, *J. Biol. Chem.*, **270**:3512–3517.

Watling, D., Guschin, D., Muller, M., Silvennoinen, O., Witthuhn, B.A., Quelle, F.W., Rogers, N.C., Schindler, C., Stark, G.R., Ihle, J.N., and Kerr, I.M., 1993, Complementation by the protein tyrosine kinase JAK2 of a mutant cell line defective in the interferon-γ signal transduction pathway, *Nature*, **366**:166–170.

Wilkes, A.F., and Harpur, A.G., 1994, Cytokine signal transduction and the JAK family of protein tyrosine kinases, *Bioessays*, , **16**:313–320.

Wilkes, A.F., Harpur, A.G., Kurban, R.R., Ralph, S.J., Zurcher, G., and Ziemiecki, A., 1991, Two novel protein-tyrosine kinases, each with a second phosphotransferase-related catalytic domain, define a new class of protein kinase, *Mol. Cell. Biol.*, **11**:2057–2065.

Witthuhn, B.A., Quelle, F.W., Silvennoinen, O., Yi, T., Tang, B., Miura, O., and Ihle, J.N., 1993, JAK2 associates with the erythropoietin receptor and is tyrosine phosphorylated and activated following stimulation with erythropoietin, *Cell*, **74**:227–236.

Witthuhn, B.A., Silvennoinen, O.M., Lai, K.S., Cwik, C., Liu, E.T., and Ihle, J.N., 1994, Involvement of the Jak-3 Janus kinase in signalling by interleukin 2 and 4 in lymphoid and myeloid cells, *Nature*, **370**:153–157.

Yin, T., Yang, L., and Yang, Y.C., 1995, Tyrosine phosphorylation and activation of JAK family tyrosine kinases by interleukin-9 in MO7E cells, *Blood*, **85**:3101–3106.

Zhao, Y., Wagner, F., Frank, S.J., and Kraft, A.S., 1995, The amino-terminal portion of the JAK2 protein kinase is necessary for binding and Phosphorylation of the granulocyte-macrophage colony-stimulating factor receptor β_c chain, *J. Biol. Chem.*, **270**:13814–13818.

Zhuang, H., Patel, S.V., He, T.C., Niu, Z., and Wojchowski, D.M., 1994, Dominant negative effects of a carboxy-truncated JAK2 mutant on EPO-induced proliferation and JAK2 activation, *Biochem. Biophys. Res. Commun.*, **204**:278–283.

8

The pp70S6 Kinase is Activated by a Complex Multi-step Process Involving Several Signaling Pathways

TIMOTHY C. GRAMMER and JOHN BLENIS

1. Introduction

Over the past several years, there has been a tremendous advance in the identification of regulatory proteins involved in the control of cellular processes such as mitogenesis, motility, differentiation, and death. However, our understanding of the way these proteins function and interact to regulate these events has not kept pace with the rate of protein discovery. A challenging but extremely important task is to understand the complex communication network of intracellular signaling pathways to which these proteins belong.

One molecule that appears crucial for cell proliferation is a serine/threonine kinase, called pp70S6k. This enzyme has been the subject of intense study, and recent work from our laboratory has helped to identify signaling molecules that play important roles in its activation. Some of that work will be discussed here, and a model will be proposed to explain its complex regulation.

2. Activators of pp70S6k

2.1. mTOR

The fungal antibiotic rapamycin has strong antiproliferative effects and completely inhibits the stimulation of pp70S6k by a wide variety of activators (Calvo *et al.*,

Tel: 617-432-1281; Fax: 617-432-1144

1992; Chung *et al.*, 1992; Kuo *et al.*, 1992; Price *et al.*, 1992; Terada *et al.*, 1982). Rapamycin binds to an intracellular receptor, FKBP, and the rapamycin:FKBP complex then exerts antiproliferative effects by targeting other intracellular proteins (Schreiber, 1991). A rapamycin:FKBP targeted protein has recently been identified (for review see Chou and Blenis, 1995; Dumont and Su, 1995), and is referred to as the mammalian target of rapamycin or mTOR (also known as FRAP or RAFT). Only an autophosphorylating kinase activity has been demonstrated for mTOR, but its similarity to the lipid kinases phosphatidylinositol-3 kinase (PI(3)K) and VPS34 suggests that it may be a lipid kinase. mTOR was recently shown to be a rapamycin-sensitive activator of pp70S6k (Brown *et al.*, 1995), although the exact mechanism for this is unclear.

2.2. Protein Kinase C

There are multiple isoforms of the serine/threonine kinase protein kinase C (PKC); these can be divided into subclasses based on their sensitivities to activators and their requirements for cofactors (Nishizuka, 1992). Early studies of pp70S6k regulation indicate a positive role for the conventional PKC (cPKC) isoforms. Phorbol esters such as phorbol-12-myristate 13-acetate (PMA) are tumor promoters that directly activate cPKC isoforms, which then rapidly induce pp70S6k activation. Inhibition of cPKCs, such as by long-term pretreatment with PMA (a process referred to as downregulation), can attenuate pp70S6k activation by many stimuli. Downregulation completely inhibits pp70S6k activation by PMA (Blenis and Erikson, 1986; Pelech and Krebs, 1987) and partially inhibits pp70S6k activation by serum, oncogenes (i.e. v-*src*), and receptor tyrosine kinases such as the epidermal growth factor (EGF) receptor and platelet-derived growth factor (PDGF) receptor. The partial inhibition of pp70S6k activation by cPKC inhibition led to the proposal that some mitogenic stimuli may utilize both cPKC-dependent and -independent pathways to stimulate pp70S6k (Blenis and Erikson, 1986; Chen *et al.*, 1991; Chung *et al.*, 1994; Pelech and Krebs, 1987; Susa *et al.*, 1989). The identity of the cPKC-independent pathway appears to involve the lipid kinase PI(3)K, as explained below.

2.3. Phosphatidylinositol-3-kinase

2.3.1. Genetic Evidence

Recently, experiments using PDGF receptor mutants have led to the identification of two distinct pp70S6k stimulating pathways. Upon ligand binding, the PDGF receptor autophosphorylates on several tyrosine residues which then serve as binding sites for downstream Src homology 2 (SH2)-containing proteins (Fantl *et al.*, 1992; Valius and Kazlauskas, 1993). Receptors whose tyrosines have been changed to phenylalanines become impaired in their ability to activate pathways requiring these phosphotyrosine sites. Such mutants were used to study PDGF-mediated signaling; they revealed two regions of the receptor that are utilized for pp70S6k stimulation (Chung *et al.*, 1994), (Fig. 1). One binding region activates phospholipase C (PLC)-gamma, which activates cPKCs. The other region binds to PI(3)K and is responsible for the majority of PDGF-dependent pp70S6k activation. This activation is cPKC-independent. These receptor sites are not required for other PDGF mediated signals, such as the activation of the MAP kinase pathway. PI(3)K is activated by many mitogens and acts as both as lipid and protein kinase (Kapeller and Cantley, 1994); however, downstream effectors of PI(3)K have only recently been discovered. The first molecular target of its action to be identified was pp70S6k.

These findings imply that the PLC-gamma/cPKC and PI(3)K signaling pathways may mediate pp70S6k activation in response to other mitogens as well. A similar receptor mutant approach was used in our laboratory to study interleukin-2 (IL-2) signaling, and we found that IL-2 dependent activation of pp70S6k also requires PI(3)K activation (Monfar *et al.*, 1995) (Fig. 1). IL-2 does not appear to activate the PLC-gamma/cPKC pathway, and consistent with this, the activation of pp70S6k is completely dependent on PI(3)K activation.

The hormone insulin has also been analyzed for its ability to activate pp70S6k. A genetic complementation approach revealed that the insulin receptor utilizes one of its major substrates, the insulin receptor substrate-1 (IRS-1), to activate pp70S6k, and that it appears to occur through the binding and activation of PI(3)K (Myers *et al.*, 1994) (Fig. 2). Once again, this stimulatory pathway is distinct from that utilized for the activation of the MAP kinase pathway. Several PI(3)K binding sites exist on IRS-1, and when these are mutated to phenylalanines, insulin-dependent activation of PI(3)K and pp70S6k is lost. An IRS-1 mutant with many of its tyrosines changed to phenylalanines, but retaining the PI(3)K binding sites, can still activate both PI(3)K and pp70S6k (Myers, M.G.Jr., Grammer, T.C., Blenis, J., and White, M.F. unpublished). Similar to IL-2, insulin does not activate the PLC-gamma/cPKC pathway, and its activation of pp70S6k can be explained solely by the PI(3)K-mediated signal. IRS-1 is essential for the activation of PI(3)K and pp70S6k, and for proliferation by the mitogens insulin-like growth factor-1 (IGF-1) and interleukin-4 (IL-4) as well (Myers *et al.*, 1994).

2.3.2. Pharmacological Evidence

Pharmacological agents have also been used to link PI(3)K to pp70S6k activation. The antibiotic wortmannin specifically inhibits PI(3)K at low nanomolar concentrations (Arcaro and Wymann, 1993; Yano *et al.*, 1993). Dose-response curves show that wortmannin inhibition of PDGF-dependent PI(3)K activity

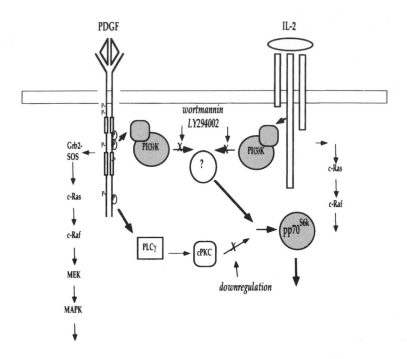

Figure 1. PDGF and IL-2 activate pp70S6k via the wortmannin/LY294002-sensitive PI(3)K pathway. This pathway is distinct from the pathway leading to the activation of MAP kinase. PDGF also utilizes the PLC-gamma/cPKC pathway to activate pp70S6k. The letter "p" on the PDGF receptor represents autophosphorylated tyrosine residues, which serve as binding sites for different signalling molecules.

parallels inhibition of pp70S6k stimulation (Chung *et al.*, 1994). PDGF-dependent activation of pp70S6k is largely sensitive to wortmannin, with the remaining activity being inhibitable by cPKC downregulation (Chung *et al.*, 1994). This observation strongly agrees with the results obtained using the PDGF receptor mutants, where pp70S6k activation occurs via PI(3)K and PLC-gamma (Fig. 1). IL-2 and insulin activation of pp70S6k is completely inhibited by wortmannin but is resistant to downregulation, which is consistent with their ability to activate PI(3)K but not cPKCs (Cheatham *et al.*, 1994; Chung *et al.*, 1994; Monfar *et al.*, 1995). PMA, a direct activator of cPKCs, activates pp70S6k in a wortmannin-insensitive, but completely downregulation-inhibitable manner (Chung *et al.*, 1994).

A second PI(3)K inhibitor, LY294002, which is structurally unrelated to wortmannin, also inhibits pp70S6k activation at doses that parallel those for

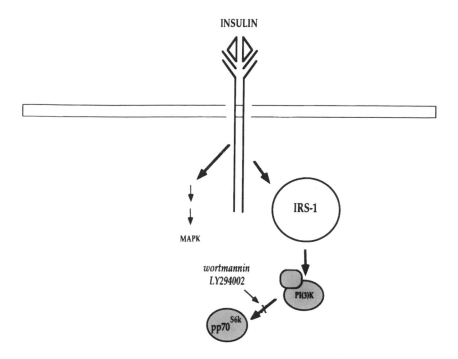

Figure 2. Insulin activates pp70S6k via the PI(3)K pathway, which requires the IRS-1 protein. This is distinct from the pathway leading to the activation of MAP kinase.

PI(3)K inhibition (Cheatham *et al.*, 1994; Vlahos *et al.*, 1994). Experiments using this inhibitor, similar to those outlined above for wortmannin, yield the same conclusion: PI(3)K activates pp70S6k (Cheatham *et al.*, 1994; Chung *et al.*, 1994; Monfar *et al.*, 1995).

2.4. Rho-family G Proteins

The Rho-family of G proteins have been linked to a variety of intracellular responses and have been shown to regulate a variety of kinases. Work from our laboratory has found that the Rho-family members Cdc42 and Rac1 are involved in the activation of pp70S6k (Chou and Blenis, 1996). This can be demonstrated in a number of ways. Activated forms of these proteins causes the activation of pp70S6k, and inhibitory isoforms block pp70S6k activation.

A particularly interesting finding is that the activated G proteins appear to bind to an inactive but partially phosphorylated form of pp70S6k (Chou and Blenis, 1996). Although this binding does not appear to directly activate the kinase, it may

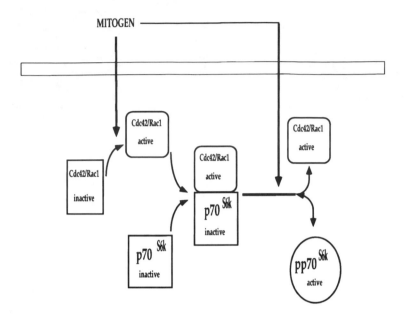

Figure 3. The active forms of the Rho-family G protein members Cdc42 and Rac1 bind to the inactive pp70S6k and are involved indirectly in its activation. p70S6k and pp70S6k denote hypo- and hyperphosphorylated forms of the kinase, respectively.

be required for localizing pp70S6k inside the cell, where it could be activated by other molecules (Fig. 3). This is highly analogous to the role played by the G protein ras in the activation of the raf kinase.

3. Structural Features of pp70S6k

3.1. Sequence Domains of pp70S6k

pp70S6k can be roughly divided into three domains: (1) the catalytic domain, (2) the amino-terminal domain (N-terminus), and (3) the carboxyl-terminal domain (C-terminus) (Fig. 4).

3.1.1. The Catalytic Domain

The catalytic domain contains the consensus motifs conserved in serine/threonine protein kinases (Hanks *et al.*, 1988). Of the five subfamilies of protein kinases that have been described, pp70S6k most closely resembles protein kinase C, cAMP-dependent protein kinase, and pp90RSK kinase.

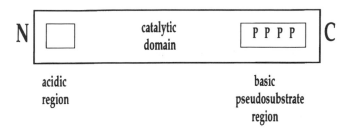

Figure 4. pp70S6k can be roughly divided into three domains: (1) the amino (N) terminus, which contains an acidic region, (2) the catalytic domain, and (3) the carboxyl (C) terminus, which contains a basic pseudosubstrate region as well as four proline-directed phosphorylation sites (denoted as P) that are the major mitogen-stimulated sites of phosphorylation.

3.1.2. The N-terminal Domain

The N-terminus contains an acidic stretch of amino acids that is hypothesized to be involved in substrate recognition, since pp70S6k's primary substrate, the ribosomal S6 protein, is highly basic. The N-terminal acidic region may also be important for stabilizing an intramolecular interaction with a C-terminal basic region of pp70S6k, and thus may play a role in the regulation of pp70S6k, as discussed below.

3.1.3. The C-terminal Domain

The C-terminus contains a sequence that is highly homologous to the region of S6 that is phosphorylated by pp70S6k. However, pp70S6k is unable to phosphorylate the C-terminal sequence, leading to the proposal that the C-terminus acts as a pseudosubtrate/autoinhibitory region by binding and occluding the catalytic site. The C-terminus also contains four major mitogen-stimulated phosphorylation sites of pp70S6k.

The characteristics of the domains led to a model that in quiescent cells, pp70S6k is maintained in an inactive conformation by the intramolecular interaction of the C-terminal pseudosubstrate domain with the catalytic domain. This confirmation may be stabilized by a further interaction between the C-terminal basic region and the N-terminal acidic region. Mitogen stimulation then causes the phosphorylation of the four C-terminal sites, which may facilitate the release of the pseudosubtrate domain from the catalytic site and subsequently lead to pp70S6k activation. This model will be referred to below, when site-specific and deletion mutant analysis experiments are outlined.

3.2. Direct Analysis of pp70S6k

3.2.1. Phosphorylation Analysis of pp70S6k

It is clear from the previous sections that multiple signaling pathways converge to activate pp70S6k. This activation of pp70S6k is accompanied by its phosphorylation, and many of the phosphorylation sites on pp70S6k have been mapped and studied for their role in pp70S6k regulation.

There are four amino acids in the pp70S6k C-terminal domain that are the major mitogenically-induced phosphorylation sites. These phosphorylated residues are immediately followed by proline residues, suggesting that proline-directed kinases are involved. Cdc2 and MAP kinases, members of the proline-directed kinase family, can phosphorylate these sites *in vitro*, but they do not activate pp70S6 (Mukhopadhayay *et al.*, 1992); (Grammer, T.C., and Blenis, J. unpublished). The more recently described stress-activated proline-directed kinases, such as p38 and Jnk, may be candidates for this phosphorylation. Indeed p38 is able to phosphorylate the C-terminus of pp70S6k *in vitro* (Grammer, T.C. and Blenis, J. unpublished). However, there is no evidence that this phosphorylation or the Cdc2/MAPK phosphorylation occurs *in vivo*, or that it activates pp70S6k.

These four sites appear to be important for full activation of pp70S6k; however (in agreement with the *in vitro* analysis mentioned), phosphorylation of these sites is not sufficient for activation, since mutating these residues to acidic, phosphorylation-mimicking residues does not cause activation of pp70S6k (Cheatham *et al.*, 1995; Ferrari *et al.*, 1993; Han, *et al.*, 1995).

Additional *in vivo* phosphorylation sites on pp70S6k have been partially characterized, and wortmannin- and rapamycin-dependent inhibition of pp70S6k is accompanied by site-specific dephosphorylation of several of these sites. It has been shown that both rapamycin and wortmannin prevent the mitogen-activated phosphorylation of the same four sites in the enzyme, three of which are different from the proline-directed sales mentioned above. At first glance, this suggests that these drugs inhibit the same pathway. Such an interpretation is unlikely, however, since a rapamycin-resistant, but wortmannin-sensitive, mutant of pp70S6k has been identified (see next section).

3.2.2. Deletion Analysis of pp70S6k

Characterization of deletion mutants of the pp70S6k has revealed further information about pp70S6k regulation (refer to Fig. 5). Deletion of the C-terminal 100

pp70S6k Construct	Mitogen Stimulated Activity	Rapamycin Sensitivity	Wortmannin Sensitivity
wild type	+++++	+++++	+++
N-deletion	+	++	+
C-deletion	++++	+++	+++++
N-/C-deletion	+++	-	+++

Figure 5. Deletion of either the amino (N-) terminus, the carboxyl (C-) terminus, or both (N-/C-) changes the responsiveness of pp70S6k to mitogens as well as to the indirect inhibitors rapamycin and wortmannin, as compared to the full-length (wild type) protein.

amino acids removes the basic pseudosubtrate domain. However, this mutant is not constitutively active, but is mitogen-regulated in a manner similar to the wild type enzyme. This is surprising, since the mutant lacks the four major mitogen-induced proline-directed phosphorylation sites (Cheatham *et al.*, 1995; Weng *et al.*, 1995). In contrast, deletion of the first 30 amino acids in the N-terminus of pp70S6k almost completely abrogates enzyme activity. Most surprisingly, when the N-terminal deletion is combined with the C-terminal deletion, the resultant enzyme regains mitogen-activated activity. The simplest interpretation of these results is that the N-terminus is needed to relieve an inhibition imposed by the C-terminus.

Analysis of the drug sensitivity of these various mutants further emphasizes the complexity of pp70S6k regulation (Cheatham *et al.*, 1995; Weng *et al.*, 1995). The N-terminal/C-terminal double mutant is rapamycin resistant, yet fully sensitive to wortmannin. This suggests that the targets of these drugs, mTOR and PI(3)K respectively, function in different pathways, and although rapamycin and wortmannin prevent the phosphorylation of an overlapping set of sites (Han *et al.*, 1995), there is probably an additional site(s) that is uniquely targeted by rapamycin. This is supported by the observation that pp70S6k from quiescent, rapamycin-treated cells migrate differently (apparently hypophosphorylated) in SDS-PAGE gels than does pp70S6k from wortmannin-treated cells (Grammer, T.C., Chou, M.M., Cheatham, L., and Blenis, J., unpublished results).

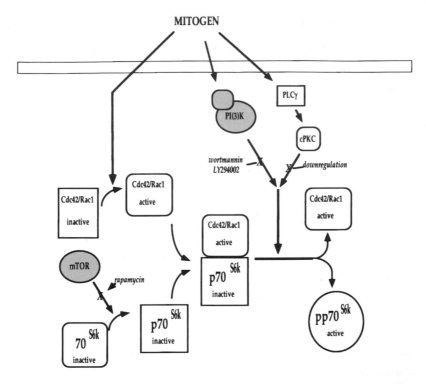

Figure 6. A model for the activation of pp70S6k is shown here as a stepwise process involving the rapamycin-sensitive mTOR activity, the binding of the active Cdc42 and Rac1 G proteins, and the wortmannin-sensitive PI(3)K and downregulation-sensitive cPKC activities. 70S6k, p70S6k, and pp70S6k denote increasing phosphorylation states of the kinase.

4. Model of pp70S6k Activation

The model of pp70S6k activation that is most consistent with the known data is as follows (Fig. 6): The rapamycin-sensitive/mTOR-dependent activity appears necessary for the phosphorylation of basally phosphorylated but not activating sites. Upon mitogen stimulation, the four proline-directed sites in the C-terminus are phosphorylated. This may allow pp70S6k to interact with the active forms of Cdc42 and Rac1, which may recruit the kinase to specific intracellular locations. Here, the enzyme is activated by the PI(3)K-dependent or the PLC-gamma/cPKC-dependent pathways.

This model can explain why rapamycin inhibits pp70S6k activation by all mitogens, since mTOR activity is required to "prime" the kinase for subsequent stimulation. The priming event may affect the intramolecular structure of pp70S6k, allowing it to be recognized by mitogen-stimulated enzymes. This priming can be circumvented by removing the N and C-termini of pp70S6k, yielding a protein that is resistant to rapamycin's effects, but still requires the activity of the wortmannin-sensitive pathway for its stimulation. The model above assumes that pp70S6k phosphatases are constitutively active and cause pp70S6k to be rapidly dephosphorylated and inactivated in the absence of mTOR activity. The finding that rapamycin and wortmannin apparently target the same phosphorylation sites can be explained by the activity of these phosphatases to remove the activating phosphorylations that occur after PI(3)K activation. We predict, however, that these activating sites may not be phosphorylated in the presence of mTOR activity alone, and that other sites of mTOR regulation remain to be identified. PI(3)K and cPKC inhibitors will have no, partial, or complete inhibitory effects on pp70S6k stimulation, depending on whether the mitogen involved utilizes the corresponding activating pathway.

5. Summary

The experiments mentioned in this report have helped to clarify the complex regulatory network that activates pp70S6k, but much more work is necessary to obtain a complete understanding of these signaling pathways. This will not be an easy task, as there appears to be a considerable amount of crosstalk and feedback loops among the individual signaling systems. The reward for accomplishing this goal will be great, however, as these pathways play crucial roles in both normal cellular physiological functions as well as abnormal and clinically relevant disease states.

References

Arcaro, A., and Wymann, M.P., 1993, Wortmannin is a potent phosphatidylinositol 3-kinase inhibitor: the role of phosphatidylinositol 3,4,5-trisphosphate in neutrophil responses, *Biochem. J.*, **296**:297–301.

Blenis, J., and Erikson, R.L., 1986, Stimulation of ribosomal S6 kinase activity by $pp60^{v-src}$, or by serum: dissociation from phorbol ester-mediated activity, *Proc. Natl. Acad. Sci. USA*, **82**:1733–1737.

Brown, E.J., Beal, P.A., Keith, C.T., Chen, J., Shin, T.B., and Schreiber, S.L., 1995, Control of p70 S6 kinase by kinase activity of FRAP *in vivo*, *Nature*, **377**:441–446.

Calvo, V., Crews, C.M., Vik, T.A., and Bierer, B.E., 1992, Interleukin 2 stimulation of p70 S6 kinase activity is inhibited by the immunosuppressant rapamycin, *Proc. Natl. Acad. Sci. USA*, **89**:426–429.

Cheatham, B., Vlahos, C.J., Cheatham, L., Wang, L., Blenis, J., and Kahn, C.R., 1994, Phosphatidylinositol 3-kinase activation is required for insulin stimulation of pp70S6 kinase, DNA synthesis, and glucose transporter translocation, *Mol. Cell. Biol.*, **14**(7):4902–4911.

Cheatham, L., Monfar, M., Chou, M., and Blenis, J., 1995, Structural and functional analysis of pp70S6k regulation, *Proc. Natl. Acad. Sci. U.S.A.*, **92**:11696–11700.

Chen, R.-H., Chung, J., and Blenis, J., 1991, Regulation of $pp90^{rsk}$, phosphorylation and S6 phosphotransferase activity in Swiss 3T3 cells by growth factor-, phorbol ester-, and cyclic AMP-mediated signal transduction, *Mol. Cell. Biol.*, **11**(4):1861–1867.

Chou, M.M., and Blenis, J., 1995, The 70 kD S6 kinase: regulation of a kinase with multiple roles in mitogenic signalling, *Current Opinions in Cell Biology*, **7**:806–814.

Chou, M.M., and Blenis, J., 1996, the 70 kD S6 kinase complexes with and is activated by the Rho family of G proteins Cdc42 and Rac1, *Cell*, **85**:573–583.

Chung, J., Grammer, T.C., Lemon, K.P., Kazlauskas, A., and Blenis, J., 1994, PDGF- and insulin-dependent pp70S6k activation mediated by phosphatidylinositol-3-OH kinase, *Nature*, **370**:71–75.

Chung, J., Kuo, C.J., Crabtree, G.R., and Blenis, J., 1992, Rapamycin-FKBP specifically blocks growth-dependent activation of and signalling by the 70 kD S6 protein kinases, *Cell*, **69**:1227–1236.

Dumont, F.J., and Su, Q., 1995, Mechanism of action of the immunosuppressant rapamycin, *Life Sciences:*, **58**(5):373–395.

Fantl, W.J., Escobedo, J.A., Martin, G.A., Turck, C.W., Rosario, M.D., McCormick, F., and Williams, L.T., 1992, Distinct phosphotyrosines on a growth factor receptor bind to specific molecules that mediate different signaling pathways, *Cell*, **69**:413–423.

Ferrari, S., Pearson, R.B., Siegmann, M., Kozman, S.C., and Thomas, G., 1993, The immunosuppressant rapamycin induces inactivation of $p70^{S6k}$, through dephosphorylation of a novel set of sites., *J. Biol. Chem.*, **268**:16091–16094.

Han, J.W., Pearson, R.B., Dennis, P.B., and Thomas, G., 1995, Rapamycin, wortmannin, and the methylxanthine SQ20006 inactivate $p70^{S6k}$, by inducing dephosphorylation of the same subset of sites, *J. Biol. Chem.*, **270**(36):21396–21403.

Hanks, S.K., Quinn, A.M., and Hunter, T., 1988, The protein kinase family: conserved features and deduced phylogeny of the catalytic domains, *Science*, **241**:42–52.

Kapeller, R., and Cantley, L.C., 1994, Phosphatidylinositol 3-kinase, *Bioessays*, **16**(8):565–576.

Kuo, C.J., Chung, J., Florentino, D.F., Flanagan, W.M., Blenis, J., and Crabtree, G.R., 1992, Rapamycin selectively inhibits interleukin-2 activation of p70 S6 kinase, *Nature*, **358**:70–73.

Monfar, M., Lemon, K.P., Grammer, T.C., Cheatham, L., Chung, J., Vlahos, C.J., and Blenis, J., 1995, Activation of pp70/85–S6 kinases in IL-2-responsive lymphoid cells is mediated by phosphatidylinositol 3-kinase and inhibited by cyclic AMP, *Mol. Cell. Biol.*, **15**:326–337.

Mukhopadhayay, N.K., Price, D.J., Kyriakis, J.M., Pelech, S., Sanghera, J., and Avruch, J., 1992, An array of insulin-activated, proline directed serine/threonine protein kinases phosphorylate the p70 S6 kinase, *J. Biol. Chem.*, **267**:3325–3335.

Myers, M.G., Grammer, T.C., Wang, L.-M., Sun, X.J., Pierce, J.H., Blenis, J., and White, M.F., 1994, Insulin receptor substrate-1 mediates phosphatidylinositol-3'-kinase and pp70S6k signalling during insulin, insulin-like growth factor-1, and interleukin-4 stimulation, *J. Biol. Chem.*, **269**:28783–28789.

Nishizuka, Y., 1992, Intracellular signalling by hydrolysis of phospholipids and activation of protein kinase C, *Science*, **258**:607–614.

Pelech, S.L., and Krebs, E.G., 1987, Mitogen-activating S6 kinase is stimulated via protein kinase C-dependent and independent pathways in Swiss 3T3 cells, *J. Biol. Chem.*, **262**:11598–11606.

Price, D.J., Grove, J.R., Calvo, V., Avruch, J., and Bierer, B.E., 1992, Rapamycin-induced inhibition of the 70-kilodalton S6 protein kinase, *Science*, **257**:973–977.

Schreiber, S.L., 1991, Chemistry and biology of the immunophilins and their immunosuppressant ligands, *Science*, **251**:283–287.

Susa, M., Olivier, A.R., Fabbro, D., and Thomas, G., 1989, EGF induces biphasic S6 kinase activation: late phase is protein kinase C-dependent and contributes to mitogenicity, *Cell*, **57**:817–824.

Terada, N., Lucas, J.J., Szepesi, A., Franklin, R.A., Takase, K., and Gelfand, E.W., 1992, Rapamycin inhibits the phosphorylation of p70 S6 kinase in IL-2 and mitogen-activated human T cells, *Biochem. Biophys. Res. Commun.*, **186**(3):1315–1321.

Valius, M., and Kazlauskas, A., 1993, Phospholipase C-g1 and phosphatidylinositol 3 kinase are the downstream mediators of the PDGF receptor's mitogenic signal, *Cell*, **73**:321–334.

Vlahos, C.J., Matter, W.F., Hui, K.Y., and Brown, R.F., 1994, A specific inhibitor of phosphatidylinositol 3-kinase,2-(4-morpholinyl)-8-phenyl-4H-1-benzopyran-4-one (LY294002), *J. Biol. Chem.*, **267**(7):5241–5248.

Weng, Q., Andrabi, K., Kozlowski, M.T., Grove, J.R., and Avruch, J., 1995, Multiple independent inputs are required for activation of the p70 S6 kinase, *Mol. Cell. Biol.*, **15**:2333–2340.

Yano, H., Satoshi, N., Kimura, K., Hanai, N., Saitoh, Y., Fukui, Y., Nonomura, Y., and Matsuda, Y., 1993, Inhibition of histamine secretion by wortmannin through the blockade of phosphatidylinositol-3-kinase in RBL-2H3 cells, *J. Biol. Chem.*, **268**(34):25846–25856.

Biochemical Dissection of Nuclear Events in Apoptosis

ATSUSHI TAKAHASHI, EMAD S. ALNEMRI,
TERESA FERNANDES-ALNEMRI, YURI A. LAZEBNIK,
ROBERT D. MOIR, ROBERT D. GOLDMAN, GUY G. POIRIER,
SCOTT H. KAUFMANN and WILLIAM C. EARNSHAW

1. Introduction

Apoptosis is a stereotyped form of cell death that was originally identified based on the highly characteristic morphology exhibited by the dying cells (Kerr *et al.*, 1972). These cells typically exhibit an overall shrinkage of the cytoplasm and nucleus, a dramatic hypercondensation of the chromatin, and the formation of membrane-enclosed blebs that are phagocytosed by surrounding cells (Wyllie *et al.*, 1980). In contrast to necrotic cell death (Wyllie *et al.*, 1980), membrane integrity is lost only very late in the process. The morphological events of apoptosis are accompanied by characteristic biochemical changes in the cell. These include cleavage of the DNA (Wyllie, 1980) and a number of specific polypeptides (Kaufmann, 1989). One striking observation about apoptosis is that no matter what the original event that triggers the apoptotic response, whether it be exposure to ionizing radiation, viral infection, removal of trophic factors, activation of specific cell surface receptors, or a host of other causes, the morphological and biochemical events that occur as cells die follow the highly recognizable pathway described above (Wyllie *et al.*, 1980).

Apoptotic death is a two-phase process. Following detection by the cell of an appropriate death stimulus, the cell enters an initial, condemned, phase (Earnshaw, 1995) that can last from hours to days. The events of the condemned phase appear to vary significantly between cell types, and with differing apoptotic stimuli. An

Correspondence: William C. Earnshaw, Tel: +44-131-650-7101; Fax: +44-131-650-7100; e-mail:bill.earnshaw@ed.ac.uk important feature of the condemned phase is that reprieve is possible. This appears to be one role of several members of the Bcl-2 family of death regulators. In fact, the response of a given cell type to the death stimulus may be substantially influenced by the pattern of Bcl-2 family members expressed.

The transition from the condemned phase to the active phase of cell death is mediated by a cell-autonomous stochastic pathway. Thus, even in highly synchronized clonal cell populations, cells die with a remarkable lack of synchrony. This aspect of apoptotic death is particularly clear in tissues, where cells undergoing apoptotic death are typically surrounded by healthy neighbors.

The active phase of cell death, which we refer to as apoptotic execution (Earnshaw, 1995), is characterized by the sequence of morphological and biochemical events described above. In contrast to the varied nature of the condemned phase, the execution phase is much more rapid and consistent between cell types and with different inductive stimuli. From the first overt appearance of morphological changes to the ultimate death and disassembly of the cell typically takes between 15 minutes and 1 hour (Wyllie *et al.*, 1980). Because of this rapid and "capricious" death of individual cells in a population, it is not possible to obtain synchronous populations of cells at a specific stage of apoptotic execution. This has posed significant problems for the biochemical analysis of apoptotic mechanisms.

Our approach to the problem has been to develop a cell-free system for the study of apoptotic execution. This system utilizes highly concentrated extracts prepared from chicken DU249 cells in the condemned phase of apoptosis (Wood and Earnshaw, 1990; Lazebnik *et al.*, 1993). The cells are condemned to apoptosis as a result of an S phase cell cycle block. However, they divide one last time before undergoing apoptosis. This is crucial for the production of our apoptotic extracts (Lazebnik *et al.*, 1993). Following release of the S phase block, cells are harvested in mitosis and lysed by a novel procedure that results in the production of extracts containing organelle-free cytoplasm at concentrations between 10 and 30 mg/ml (Wood and Earnshaw, 1990). These S/M extracts rapidly and irrevocably induce the morphological and biochemical events of apoptosis in cell nuclei that have been isolated from healthy growing cells (Lazebnik *et al.*, 1993).

The S/M extract system has several advantages for the biochemical analysis of apoptotic execution. First, the added nuclei undergo every known aspect of nuclear apoptosis thus far detected in live cell systems. Second, they do so with an extraordinary degree of synchrony: they undergo a series of highly predictable minute-by-minute biochemical changes. Third, since the extracts are cell-free, they can readily be used for the characterization of enzyme activities in apoptosis, as substrates and inhibitors may be added directly. Finally, use of the extracts minimizes the possibility of indirect effects resulting from the induction of complex metabolic and signaling pathways in living cells.

The most important results obtained to date with the extracts have involved characterization of the role of ICE-related proteases (IRPs) in apoptosis, and in particular, the identification of endogenous cellular substrates for these proteases in dying cells. ICE (the interleukin-1β-converting enzyme) is an unusual cysteine protease that cleaves pro-interleukin-1β at two specific sites, both immediately adjacent to aspartate residues (Sleath *et al.*, 1990; Howard *et al.*, 1991). The realization that IRPs have an important role in cell death came with the discovery that the product of the nematode *Caenorhabditis elegans ced-3* gene exhibits 28% identity to ICE (Yuan *et al.*, 1993). *Ced-3* had been identified as one of two genes that are absolutely required for all of the 131 programmed cell deaths that occur during development of the *C. elegans* hemaphrodite (Yuan and Horvitz, 1990). The recognition of the similarity of the *CED-3* protein to ICE immediately suggested that *CED-3* initiates cell death by cleaving one or more critical target proteins. Support for this hypothesis was obtained when it was shown that constructs expressing a transfected ICE cDNA exhibited prominent features of apoptotic death (Miura *et al.*, 1993).

The genetic studies in *C. elegans* were extraordinarily fruitful in that they revealed the involvement of IRPS in apoptosis. Unfortunately, they failed to identify any substrates for the IRPs during cell death. It is here that biochemical studies with the S/M extracts have proven to be most useful. Not only have we been able to identify five specific substrates for IRPs in apoptosis, we have been able to show that these substrates are cleaved by at least two, and possibly five or more distinct IRPs acting in concert (Lazebnik *et al.*, 1994, 1995; Takahashi *et al.*, 1996).

2. Results and Discussion

2.1. Cleavage of PARP in S/M Extracts

In the course of an examination of the events of apoptotic execution in HL-60 leukemia cells, we noted that poly(ADP-ribose) polymerase, a nuclear enzyme that plays an as-yet unknown role in DNA repair, is cleaved from its normal M_r 116 kDa form to a fragment of M_r 85 kDa (Kaufmann *et al.*, 1993). This fragment is stable for hours in apoptotic cells.

After determining that both endogenous nuclear human PARP and purified bovine PARP are cleaved very rapidly in S/M extracts, we went on to map the site of PARP cleavage, and showed that cleavage occurs immediately adjacent to an asp residue in the sequence DEVD⇓G (Lazebnik *et al.*, 1994). The cleavage activity has a number of characteristics that are reminiscent of the activity of ICE: there is an absolute requirement for the presence of asp adjacent to the cleavage site, and highly specific inhibitors of ICE block the cleavage activity. However, the activity in S/M extracts does not cleave human pro-IL-1β, and purified human ICE at near physiological levels does not cleave PARP. This suggested to us that PARP is cleaved by an IRP in the chicken extract, which we designated prICE [protease resembling ICE] (Lazebnik *et al.*, 1994).

2.2. Why Cleave PARP in Apoptosis?

We do not believe that cleavage of PARP is responsible for triggering the apoptotic response. Rather, we suggest that PARP is cleaved in order to conserve cellular ATP and NAD levels, thereby maintaining the physiological environment necessary for the completion of apoptosis. This hypothesis is attractive, since in an environment where there are many DNA breaks, PARP can act as an engine for rapidly depleting cellular ATP and NAD stores. Mammalian nuclei typically contain about $10^5 - 10^6$ copies of PARP (Yamanaka *et al.*, 1988; Lindahl *et al.*, 1995). This enzyme has a low basal activity, but is dramatically activated in the presence of DNA breaks (which are generated in huge amounts during apoptotic execution). The amino-terminal DNA-binding domain of PARP binds to DNA breaks with nanomolar affinity, and this activates the enzyme (Hengartner *et al.*, 1991). Each activated enzyme rapidly automodifies itself by the addition of chains of roughly 10-200 ADP-ribose monomers (Lautier *et al.*, 1993; Lindahl *et al.*, 1995). This activation of PARP is accompanied by concomitant polymer hydrolysis by a highly active glycohydrolase, with the net result that the half-life of the ADP-ribose polymer is roughly 15–30 seconds (Alvarez Gonzalez and Althaus, 1989; Lautier *et al.*, 1993; Satoh *et al.*, 1994). With PARP working at optimum efficiency, the enzyme could consume roughly 4% of the cellular NAD per minute.

Cleavage of PARP by IRPs effectively shuts off the PARP engine by uncoupling the DNA-binding domain from the catalytic domain. Thus, while the catalytic domain remains functionally intact, it can no longer be activated by DNA breaks (Kaufmann *et al.*, 1993). We predict that if PARP cleavage did not occur, condemned phase cells would die by a necrotic pathway, thereby causing inflammation. This hypothesis is currently being tested using cell lines expressing a mutant form of PARP that cannot be cleaved by IRPs.

2.3. Cleavage of PARP by Cloned IRPs

PARP is an excellent substrate for a number of cloned human IRPs. The enzyme can be cleaved with high efficiency by CPP32 (Fernandes-Alnemri *et al.*, 1995; Nicholson *et al.*, 1995; Tewari *et al.*, 1995) and Mch3α (Fernandes-Alnemri *et al.*, 1995), and it is also cleaved, albeit less efficiently, by Mch2α (Fernandes-Alnemri *et al.*, 1995). All of these enzymes recognize predominantly the cleavage site identified *in vivo* and in S/M extracts. In addition, non-physiological levels of ICE and TX can cleave PARP (Gu *et al.*, 1995), although the biological relevance of this observation is unclear.

Although it has been claimed, based primarily on kinetic and inhibitor studies *in vitro*, that CPP32 is the *sole* PARP cleaving enzyme *in vivo* (Nicholson *et al.*, 1995), we are skeptical of this claim for two reasons. First, cloned Mch3 appears to be as active as CPP32 at PARP cleavage *in vitro* (Fernandes-Alnemri *et al.*, 1995). Thus Mch3 or other as-yet uncharacterized IRPs may participate in PARP cleavage *in vivo*. Second, the PARP protease in S/M extracts and CPP32 exhibit significant differences in their sensitivities to inhibitors. Cleavage of PARP by CPP32 expressed in *E. coli* is inhibited by TPCK, which has no effect on PARP cleavage in S/M extracts (Fernandes-Alnemri *et al.*, 1995). Thus Mch3 or other as-yet uncharacterized IRPs may participate in PARP cleavage *in vivo*. Furthermore, cloned CPP32 expressed in *E. coli* is completely inhibited by 0.1 μM YVAD-cmk, whereas up to 100 μM YVAD-cmk is required to abolish PARP cleavage in some S/M extracts (Lazebnik *et al.*, 1994, 1995). These functional differences should not be overinterpreted, as it is possible that they reflect different sensitivities of the chicken and human CPP32 enzymes. However, they certainly support the view that CPP32 may not be the sole PARP protease in vertebrates.

2.4. What is the PARP-cleaving IRP in Apoptotic Cells? Serious Problems with Current Paradigms

The fact that several IRPs can cleave PARP raises the important question: Which of these enzymes is actually responsible for PARP cleavage in apoptotic cells? This question cannot currently be answered with available technology.

At present the "gold standard" for demonstrating that IRP-x is responsible for cleaving a particular substrate is to transfect cells with a cDNA construct expressing IRP-x and demonstrate that the substrate is cleaved (Miura *et al.*, 1993). A further control for the involvement of IRPs in cleavage of a substrate is to transfect cells with a construct expressing the viral cytokine response modifier A (*crmA*) gene and demonstrate that this prevents cleavage of the substrate in question (Boudreau *et al.*, 1994; Gagliardini *et al.*, 1994). *CrmA* is known to be a specific inhibitor of some, but not all, IRPs (Ray *et al.*, 1992).

Both of these experiments suffer from important flaws. It has recently been shown that the forced introduction of a number of extracellular proteases into cells results in apoptotic death (Williams and Henkart, 1994). The proteases presumably cleave one or more essential cellular proteins, and this triggers the endogenous death pathway. This result therefore raises the possibility that any particular IRP that is transfected into cultured cells might cause cell death by a similar mechanism. Demonstrating that mutations of the active site of the transfected protease no longer cause death does not rule out this indirect mechanism, as mutant proteases would not be expected to cause intracellular damage and activate the endogenous death response. Furthermore, showing the *crmA* blocks death is likewise inconclusive, as one or more IRPs is involved in the endogenous death response (Gagliardini *et al.*, 1994), and *crmA* could therefore block events downstream of the action of the transfected protease.

A correlate of this argument shows why *crmA* inhibition of the cleavage of a particular substrate during apoptosis cannot be used to conclude that the substrate is cleaved by an IRP (Tewari *et al.*, 1995). If *crmA* blocks the endogenous death pathway, then all proteins cleaved in that pathway will be saved from cleavage. This applies equally well to the substrates for *crmA*-sensitive IRPs and to the substrates of other proteases that are activated downstream of the action of the IRPs.

At present, the best evidence for cleavage of a substrate by an IRP is the demonstration that the cloned IRP will cleave the substrate, the determination of the cleavage site, which must be adjacent to an asp, and the demonstration that the site of cleavage in cells undergoing apoptosis is the same. This approach can be strengthened when combined with the use of biotinylated affinity reagents for the direct labeling of active IRPs (see below).

2.5. Lamin Cleavage in Apoptosis

Lamin B is cleaved in HL-60 cells undergoing drug-induced apoptosis (Kaufmann, 1989), and lamins A and C are cleaved in fibroblasts undergoing apoptosis induced by exposure to specific cytotoxic T lymphocytes (Ucker *et al.*, 1992). The cleaved lamin fragments appeared to be solubilized from nuclei concomitant with cleavage (Ucker *et al.*, 1992). This was interpreted as indicative of nuclear envelope disassembly. However, it is well established that the nuclear envelope does not disassemble during apoptosis (Wyllie *et al.*, 1980). Thus lamin cleavage and solubilization in apoptosis has different functional consequences from lamin phosphorylation and solubilization in mitosis.

Studies from our laboratory have shown that lamin cleavage is carried out by a specific IRP that is distinct from the PARP protease. We showed that both endogenous nuclear lamins, as well as purified recombinant lamins A and B, are cleaved during apoptotic execution in S/M extracts (Fig. 1). The first evidence that the PARP and lamin proteases are distinct came from inhibitor studies, where it was found that TLCK blocks lamin cleavage in S/M extracts (Lazebnik *et al.*, 1995), while having no effect on PARP cleavage. Other inhibitors, including TPCK, E-64, PMSF, aprotinin, leupeptin, antipain, DCl and AAPF-cmk had no effect on lamin cleavage.

Interestingly, YVAD-cmk at 1 μM blocks lamin cleavage in S/M extracts (Lazebnik *et al.*, 1995). In the same experiments, PARP cleavage was completely inhibited only when the concentration of YVAD-cmk was raised to 100 μM (Lazebnik *et al.*, 1995). After association of the YVAD moiety with the substrate binding pocket of IRPs, the chloromethylketone rapidly alkylates the catalytic site of the larger subunit, thereby inactivating the enzyme (Walker *et al.*, 1994;

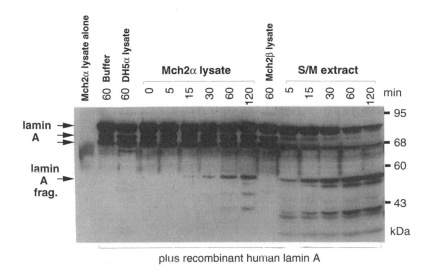

Figure 1. Cleavage of purified recombinant lamin A by cloned Mch2α and S/M Extracts. The figure shows a parallel time course of digestion of recombinant lamin A. (For experimental conditions see Takahashi *et al.*, 1996). Control lanes show that the anti-lamin antibodies do not react with proteins from the *E. coli* extracts alone (Mch2α lysate alone), and that lamin A is not cleaved in buffer, lysate from untransfected *E. coli* (DH5α lysate), or lysate from *E. coli* transfected with a catalytically inactive splice variant of Mch2 (Mch2β lysate).

Wilson *et al.*, 1994). For a given enzyme, there can be only one binding constant for the YVAD moiety, and this will inactivate the enzyme equally against all substrates. Therefore, if the PARP and lamin proteases have different sensitivities to YVAD-cmk, they must be dependent on the activities of two distinct IRPs. Our experiments reveal that the lamin protease is either an IRP or that it functions downstream of an essential IRP. Our further studies support the former hypothesis.

2.6. Lamin Cleavage by Cloned Mch2α

Proof that the lamins can be cleaved by an IRP came with the demonstration that human lamins A and B are cleaved by cloned McH2α expressed in *E. coli* (Figs. 1 and 2) (Takahashi *et al.*, 1996). In parallel experiments, CPP32 and Mch3α preparations that were highly active in PARP cleavage had no effect on lamin A (Fernandes-Alnemri *et al.*, 1995; Takahashi *et al.*, 1996). This has provided the first convincing demonstration for substrate selectivity among IRPs that are active in apoptosis. It is worth noting that ICE also failed to cleave lamin A in similar experiments.

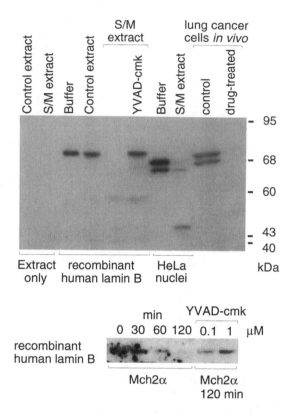

Figure 2. Cleavage of endogenous nuclear and purified recombinant lamin B during apoptosis *in vivo*, in S/M extracts and by cloned Mch2α. Upper panel: The two right lanes show that induction of apoptosis in lung cancer cells results in the disappearance of lamins B_1 and B_2. The two adjacent lanes show that lamins B_1 and B_2 are also cleaved in HeLa nuclei following addition to S/M extract. Lanes 3-6 show that purified recombinant lamin B, which is stable in buffer and control non-apoptotic extracts, is cleaved in S/M extracts by an enzyme that is senstive to YVAD-cmk. These experiments show that the lamin protease is soluble in S/M extracts, and that it shows an inhibitor sensitivity characteristic of IRPs. The two left lanes show that control and S/M extracts do not contain antigens that react with the anti-lamin antibody. Lower panel: Purified recombinant lamin B is degraded in an *E. coli* lysate containing Mch2α. This degradation is sensitive to YVAD-cmk, consistent that the fact that Mch2α is an IRP.

The activity of Mch2α against lamins is particularly notable, as this represents the cleavage of not just one polypeptide, but the products of three distinct genes (lamins A and C arise by alternative splicing from a single gene, but lamins B1 and B2 are the products of distinct genes). Thus, Mch2α appears to have become specialized for lamin cleavage (although, of course, the enzyme may cleave other substrates as well).

2.7. The Site of Lamin A Cleavage

Initial observations that lamin A is cleaved to a ~45–50 kDa fragment were not particularly surprising, as this corresponds fairly well to the length of the lamin rod domain. In general, proteolysis of intermediate filament proteins initially yields a fragment of roughly this size, reflecting the relative insensitivity of the coiled-coil domain to attack by proteases. However, our studies of lamin A cleavage in apoptosis suggested that the apoptotic protease cleaves lamin A within the rod domain. This was based on the fact that an antibody directed against a peptide at the exact carboxy-terminus of lamin A recognized a 45 kDa fragment following lamin cleavage in S/M extracts (Lazebnik *et al.*, 1995).

Analysis of the primary sequences of the cloned lamins from several species revealed that asp 230 is relatively conserved, that this residue lies in a context that is similar to the PARP cleavage site (DEVD⇓G in PARP, VEID⇓N in lamin A), and that cleavage at this site would produce a carboxy-terminal fragment of the observed size. We therefore predicted that this would be the site of lamin A cleavage. In order to determine the lamin A cleavage site, we expressed a lamin A-derived fragment (residues 1–463) containing a carboxy-terminal histidine tag in *E. coli*, isolated this protein on a Ni column, and added it to bacterial extracts containing active Mch2α (Takahashi *et al.*, 1996). Following a repetition of the Ni chromatography, several polypeptide species of the expected molecular weight were observed in SDS-PAGE. Three of the bands were excised and subjected to amino acid sequence determination. Two of the bands yielded irrelevant sequences. The third yielded 3–5 amino acids at each position. Following subtraction of perfect 12/12 matches to the *E. coli* 30 kD and curved DNA binding proteins, we were left with a partial sequence containing 9 of 12 residues with an exact match to the sequence of lamin A at the cleavage site predicted above.

We next synthesized an 11-mer peptide corresponding to the observed cleavage site. Addition of this peptide to S/M extracts significantly lowered the level of lamin proteolysis (Takahashi *et al.*, 1996). A control peptide in which the crucial asp was mutated to ala had no effect on lamin cleavage. This experiment confirms that the lamin protease in S/M extracts and Mch2α both recognize the same cleavage site on lamin A. Thus the lamin protease in apoptosis has an unusual specificity, in that its preferred cleavage site lies within the rod domain. This cleavage would be expected to divide the lamin A dimers (Moir *et al.*, 1991) into head and tail fragments and therefore disrupt both their longitudinal association into head-to-tail polymers (Heitlinger *et al.*, 1992) and their further assembly into a filamentous structure.

2.8. The Importance of Lamin Cleavage in Apoptosis

Lamin cleavage appears to be required for the completion of the disassembly of the nucleus into apoptotic bodies. If lamin cleavage is blocked by TLCK treatment, then the process of nuclear disassembly halts at the stage where the chromatin has collapsed outwards against the nuclear rim, and before the chromatin rim separates into discrete masses (Lazebnik et al., 1995). The most likely explanation for this result is that following the collapse of the chromatin against the nuclear periphery, the chromatin remains tethered against the filamentous network of the lamins through protein: DNA (Ludérus et al., 1992) and protein: protein interactions (Glass et al., 1993; Taniura et al., 1995). The latter interactions may both involve binding of the carboxy-terminal tail regions of the lamins to histones (Taniura et al., 1995), and interactions of the lamin rod domains with as-yet unidentified chromosomal proteins (Glass et al., 1993). Cleavage of the lamin rod domain by an IRP might disrupt this lamin: chromatin interaction, and release the chromatin to condense into individual domains. Cleavage of lamins may also disrupt their interaction with other nuclear components such as the retinoblastoma gene product (Ozaki et al., 1994) and adenovirus E1B 19 kDa-protein (Farrow et al., 1995). The significance of these other protein: protein interactions in the disintegration of apoptotic nuclei remains to be determined.

2.9. Does Inhibition of Specific IRPs by Zn^{2+} Explain the Protective Effects of Zn^{2+} Against Apoptosis?

Lamin cleavage by Mch2α and by S/M extracts is sensitive to millimolar concentrations of Zn^{2+} (Takahashi et al., 1996). In contrast, cleavage of PARP by CPP32 and in S/M extracts is resistant to Zn^{2+} (Fig. 3). Even though the concentrations of Zn^{2+} required for the inhibition of the lamin-cleaving IRP appear to exceed physiological levels, it is possible that compartmentalization of Zn^{2+} by binding to intracellular proteins (Pattison and Cousins, 1986) might generate sufficient concentrations in the vicinity of IRPs.

These observations are particular interesting in view of previous observations suggesting that intracellular Zn^{2+} can modulate apoptosis (Sundeman, 1995). On the one hand, Zn^{2+} depletion has been associated with enhanced apoptosis in the small intestine (Elmes, 1977) and thymus (Fraker et al., 1977) of rodents as well as with induction of apoptosis in cultured cells (Martin et al., 1991; Zalewski et al., 1991; McCabe et al., 1993; Treves et al., 1994). On the other hand, Zn^{2+} supplementation inhibits internucleosomal DNA fragmentation and morphological features of apoptosis in vivo (Waring et al., 1990; Takano et al., 1991; Zalewski et al., 1991; Cohen et al., 1992; Bicknell et al., 1994) and in vivo (Lazebnik et al., 1993; Newmeyer et al., 1994). The target for this suppression of apoptosis by Zn^{2+} remains unclear, although apoptotic nucleases have been suggested as possible candidates (Duke et al., 1983; Cohen and Duke, 1984). However, these nucleases are activated only after the initiation of apoptotic execution. It has also recently been reported that Zn^{2+} inhibits apoptosis-associated serine-protease activity in hepatocytes (Kwo et al., 1995), although the role of these proteases in apoptotic

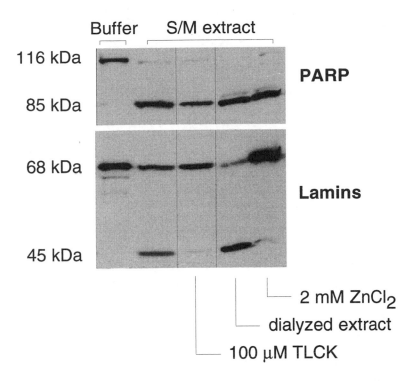

Figure 3. Zn^{2+} inhibition of lamin cleavage by S/M extracts with no effect on PARP cleavage. The two left lanes show the cleavage of PARP and lamin A in HeLa nuclei by S/M extracts [both panels come from the same gel probed as described (Lazebnik *et al.*, 1995)]. Lamin A, but not PARP, cleavage is sensitive to TLCK (center lane). The right two lanes show that S/M extract dialyzed into a buffer lacking EGTA is still active in PARP and lamin A cleavage, but that lamin cleavage by this dialyzed extract is abolished by 2 mM Zn^{2+}. This concentration of Zn^{2+} has no effect on PARP cleavage in the same extract.

death is currently unknown. The ability of Zn^{2+} to rescue some cell types from death (Sunderman, 1995) suggests that important target(s) for Zn^{2+} may act at the head of the death pathway.

In some cells, such as glucocorticoid- or etoposide-treated thymocytes (Cohen *et al.*, 1992; Sun *et al.*, 1994), Zn^{2+} blocks the later phases of apoptosis but not the initiation of apoptotic execution or ultimate cell death. We suggest that the initiation of apoptotic execution in such cells may be driven by a Zn^{2+}-insensitive IRP similar to CPP32. Other types of cells undergoing apoptosis, such as U937 cells treated with inhibitors of macromolecular synthesis (Bicknell *et al.*, 1994), are completely rescued by Zn^{2+}, presumably because apoptotic execution is initiated by means of a Zn^{2+}-sensitive IRP similar to Mch2α. Variations of sensitivity of

IRPs to Zn^{2+}, coupled with variations in expression of IRPs in different tissues, might help to explain why Zn^{2+} has different effects in different cells undergoing apoptosis.

2.10. Apoptosis in Vertebrate Cells: A Tale of many IRPs

A genetic analysis in C. elegans has thus far yielded a single IRP, the CED-3 protein, that is required for developmental cell deaths (Ellis et al., 1991). The situation in humans is very different. PCR cloning experiments have thus far yielded a total of ten published IRPs, with at least four further examples still to be published. We have recently performed a phylogenetic analysis on the small subunit of the human IRPs, based on a combination of known and deduced processing sites of the various cloned precursors. The small subunit was chosen for this analysis, as it is likely to have an important role in determining substrate specificity (Walker et al., 1994; Wilson et al., 1994). Our analysis suggested that the published IRPs fall into three groupings. The first contains the most ICE-like IRPs, ICE (Cerretti et al., 1992; Thomberry et al., 1992), TX [$ICE_{rel}II$, Ich-2] (Faucheu et al., 1995; Kamens et al., 1995; Munday et al., 1995), and $ICE_{rel}III$ (Munday et al., 1995). The second group contains the most CED-3-like human IRPs, CPP32 [Yama] (Fernandes-Alnemri et al., 1994; Tewari et al., 1995) and Mch2 (Fernandes-Alnemri et al., 1995). The third group has a single member, Ich-1 [originally described in the mouse as Nedd-2] (Kumar et al., 1994; Wang et al., 1994).

This analysis only begins to reveal the complexity of the situation, however, as most or all of these genes are processed into alternatively spliced variants, some of which act as dominant negative functional inhibitors. For example, ICE-ε, an alternatively-spliced product from the ICE gene, interferes with the induction of apoptosis by active ICE isoforms (Alnemri et al., 1995). A similar inhibitory effect has been noted for ICH-1$_s$, an alternatively spliced form of the Ich-1 transcription that interferes with the induction of apoptosis by ICH-1 (Wang et al., 1994). Short forms have also been noted for other IRPs.

Given the multiplicity of IRPs and the additional complexity arising from alternative processing of the transcripts, how are we to understand the role of particular IRPs in apoptotic events? The obvious strategy is to take a genetic approach and create mice null for each IRP in turn. Thus far, knockout mice are available only for ICE (Kuida et al., 1995; Li et al., 1995). Interestingly, these mice undergo virtually all forms of apoptosis normally, suggesting that either ICE is not involved in apoptosis, or that there is significant redundancy in the cellular repertoire of IRPs. Only cell killing by the FAS signaling pathway appears to be substantially defective in the ICE-null mice (Kuida et al., 1995).

This approach has obvious advantages, but it also has difficulties. First, the knockouts may be lethal if the IRPs have roles in pathways that are essential for life. A recent publication indicates that CPP32 may have an important role in the regulation of sterol metabolism (Wang *et al.*, 1995). If other IRPs have as-yet undiscovered roles in other essential cellular pathways, then gene knockouts may be lethal for reasons other than their effects on apoptosis. A further difficulty is presented by the presence of multiple IRP genes in humans. Given the ability of these enzymes to recognize a range of substrates, it is possible that this genetic redundancy will greatly complicate the elucidation of the IRPs responsible for specific apoptotic events *in vivo*.

We have taken a novel approach in order to address this problem. In their studies of ICE activation, the group at Merck developed a reagent that enabled them to directly label the large subunit of active ICE (Thornberry *et al.*, 1994). This reagent, N-(Acetyltyrosinylvalinyl-N^ε-biotinyllysyl) aspartic Acid [(2,6-Dimethylbenzoyl)oxy]methyl ketone, which we term YV(bio)KD-aomk, has turned out to be an extremely powerful tool for the detection of active IRPs. Like YVAD-cmk, YV(bio)KD-aomk binds into the substrate-binding pocket of active IRPs and then irreversibly alkylates the enzyme. However, the novel feature of YV(bio)KD-aomk is that the alkylated enzyme is at the same time biotinylated, and can therefore be detected with streptavidin or avidin-based reagents.

The use of this reagent directly reveals the complexity of the spectrum of active IRPs in S/M extracts (Takahashi *et al.*, 1996). When YV(bio)KD-aomk is added to the extracts at 100 μM and allowed to react with the active IRPs, and the mixture is then analyzed by SDS-PAGE and immunoblotting, at least five labeled polypeptides can be detected. We term these $prICE_{1-5}$ (Fig. 4). The bands labeled in Fig. 4 each correspond to the large subunit of an active IRP. When combined with the results of cloning experiments, where it has been found that up to eight different IRP genes are expressed in a single cell type, this experiment gives some hint as to the complexity of the spectrum of IRPs active in apoptosis.

Are the labeled bands in Fig. 4 indicative of different functional enzyme activities, or do they represent multiple isoforms of a single IRP? In the absence of specific antibody reagents, this point is difficult to prove, but our initial observations suggest that the different bands correspond to functionally distinct enzymes. The evidence for this comes from a competition experiment in which a peptide corresponding to the PARP cleavage site was added in excess to the S/M extract prior to the addition of YV(bio)KD-aomk. Despite the fact that this experiment involves competition between a reversible substrate analogue and an irreversible inhibitor, an unambiguous result was nonetheless obtained: the PARP peptide greatly lowered the labeling of $prICE_1$ and had no effect on the labeling of $prICE_{2-5}$ (Takahashi *et al.*, 1996). This is strongly suggestive evidence that $prICE_1$ is a PARP cleaving enzyme. A similar experiment with the lamin A cleavage site peptide suggested that $prICE_5$ is likely to be the lamin protease (manuscript in preparation). Again, this peptide had no effect on the labeling of $prICE_{1-4}$. This provided further evidence that the lamin protease is a soluble IRP in S/M extracts.

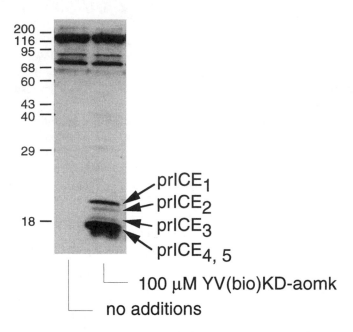

Figure 4. Direct labeling of multiple active IRPs in apoptotic S/M extracts using YV(bio)KD-aomk. S/M extract was boiled in SDS-sample buffer either directly (left lane) or following pre incubation with YV(bio)KD-aomk (right lane), subjected to SDS-PAGE and immunoblotting, and then probed with streptavidin: peroxidase. The high molecular weight proteins present in both lanes are endogenous streptavidin-binding proteins present in the extracts. prICE$_{1-5}$ correspond to the large subunits of active IRPs.

3. Conclusions

Experiments with S/M extracts have made an important contribution by identifying the first major substrates for IRPs in apoptosis. It is likely that a number of other substrates for these enzymes will be identified over the next few years. One challenge for this work will be to develop a strategy for the identification of those substrates whose cleavage is responsible for the morphological transformation of the cell during apoptotic execution. This may be difficult, as many of these subtrates may be very minor cellular constituents – for example, other enzymes that act downstream of the IRPs. However, given the great strides that have been made in cloning the many human IRPs and the possibilities that this opens up for dissection of the first steps in the apoptotic cascade, we are confident that this aspect of apoptosis research will continue to yield important new insights into apoptotic mechanisms over the next several years.

Acknowledgments

We thank Susan Molineaux and Nancy Thornberry (Merck) for YV(bio)KD-aomk and purified recombinant ICE; and Brian Burke (Calgary), Erich Nigg (Lausanne), and Larry Gerace (La Jolla) for anti-lamin antibodies. Supported by an NIH grant to SK and WCE; a Leukemia Scholar Award to SHK; and grants from the Canadian MRC and the National Cancer Institute of Canada to GPP.

References

Alnemri, E.S., Fernandes-Alnemri T., and Litwack, G., 1995, Cloning and expression of four novel isoforms of human interleukin-1-β converting enzyme with different apoptotic activities, *J. Biol. Chem.*, **270**:4312–4317.

Alvarez Gonzalez, R. and Althaus, F.R., 1989, Poly(ADP-ribose) catabolism in mammalian cells exposed to DNA-damaging agents, *Mutation Res.*, **218**:67–74.

Bicknell, G.R., Snowdon, R.T., and Cohen, G.M., 1994, Formation of high molecular mass DNA fragments is a marker of apoptosis in the human leukemic cell line, U937, *J. Cell. Sci.*, **107**:2483–2489.

Boudreau, N., Sympson, C.J., Wern, Z. and Bissell, M.J., 1994, Suppression of ICE and apoptosis in mammary epithelial cells by extracellular matrix, *Science*, **267**:891–893.

Cerretti, D.C., Kozlosky, C.J., Mosley, B., Nelson, N., van Ness,K., Greenstreet, T.A., March, C.J., Kronheim, S.R., Druck, T., Cannizzaro, L.A., Huebner, K. and Black, R.A., 1992, Molecular cloning of the interleukin-1β converting enzyme, *Science*, **256**:97–100.

Cohen, G.m., Sun, X.-M., Snowden, R.T., Dinsdale, D. and Skilleter, D.N., 1992, Key morphological features of apoptosis may occur in the absence of internucleosomal DNA fragmentation, *Biochem. J.*, **286**:331–334.

Cohen, J.J. and Duke, R.C., 1984, Glucocorticoid activation of a calcium-dependent endonuclease in thymocyte nuclei leads to cell death, *J. Immunol.*, **132**:38–42.

Duke, R.C., Chervenak, R. and Cohen, J.J., 1983, Endogenous endonuclease-induced DNA fragmentation: an early event in cell mediated cytolysis, *Proc. Nat. Acad. Sci. (USA)*, **80**:6361–6365.

Earnshaw, W.C., 1995, Apoptosis: Lessons from *in vitro* systems, *Trends Cell Biol.*, **5**:217–220.

Ellis, R.E., Yuan, J. and Horvitz, H.R., 1991, Mechanisms and functions of cell death, *Annu. Rev. Cell Biol.*, **7**:663–698.

Elmes, M.E., 1977, Apoptosis in the small intestine of zinc-deficient and fasted rats, *J. Pathol*, **123**:219–223.

Farrow, S.N., White, J., Martinou, I., Raven, T., Pun, K.T., Grinham, C.J., Martinou, J.C. and Brown, R., 1995, Cloning of a bcl-2 homologue by interaction with adenovirus e1b 19k, *Nature*, **374**:731–733.

Faucheu, C.A., Diu A., Chan, A.W.E., Blanchet, A.-M., Miossec, C., Hervé, F., Collard-Dutilleul, V., Gu, Y., Aldape, R., Lippke, J., Rocher, C., Su, M., Livingston, D., Hercend, T. and Lalanne, J.-L., 1995, A novel human protease similar to the interleukin-1β converting enzyme (ICE) induces apoptosis in transfected cells, *The EMBO J.*, **14**:1914–1922.

Fernandes-Alnemri, T., Litwack, G. and Alnemri, E.S., 1994, CPP32, a novel human apoptotic protein with homology to *Caenorhabditis elegans* cell death protein Ced-3 and mammalian interleukin-1β-converting enzyme, *J. Biol. Chem.*, **269**:30761–30764.

Fernandes-Alnemri, T., Litwack, G. and Alnemri, E.S., 1995, Mch2, a new member of the apoptotic Ced-3/Ice cysteine protease family, *Cancer Res*, **55**:2737–2742.

Fernandes-Alnemri, T., Takahashi, A., Wang, L., Yu, Z., Armstrong, B., Kribs, J., Tomasselli, K., Croce, C.M., Earnshaw, W.C., Litwack, G. and Alnemri, E.S., 1995, *Mch3*, a novel human apoptotic cysteine protease related to CPP32 and *CED-3,Cancer Res.* (in press).

Fraker, P.J., Haas, S.M., and Luecke, R.W., 1977, Effect of zinc deficiency on the immune response of the young adult A/J mouse, *J. Nutr.*, **107**:1889–1895.

Gagliardini, V., Fernandez, P.A., Lee, R.K., Drexler, H.C., Rotello, R.J., Fishman, M.C. and Yuan, J., 1994, Prevention of vertebrate neuronal death by the *crmA* gene, *Science*, **263**:826–8.

Glass, C.A., Glass, J.R., Taniura, H., Hasel, K.W., Blevitt, J.M., and Gerace, L., 1993, The α-helical rod domain of human lamins A and C contains a chromatin binding site, *The EMBO J.*, **12**:4413–4424.

Gu, Y., Sarnecki, C., Aldape, R.A., Livingston, D. and Su, M., 1995, Cleavage of poly(ADP-ribose) polymerase by interleukin-1β converting enzyme and its homologs TX and Nedd-2, *J. Biol. Chem.*, **270**:18715–18718.

Heitlinger, E., Peter, M., Lustig, A., Villiger, W., Nigg, E.A.,and Aebi, U., 1992, The role of the head and tail domain in lamin structure and assembly: analysis of bacterially expressed chicken lamin A and truncated B2 lamins, *J. Struct. Biol.*, **108**:74–89.

Hengartner, C., Lagueux, J. and Poirier, G.G., 1991, Analysis of the activation of poly(ADP-ribose)polymerase by various types of DNA, *Biochem. Cell. Biol.*, **69**:577–580.

Howard, A.D., Kostura, M.J., Thornberry, N.A., Ding, G.J.-F., Limjuco, G., Weidner, J.R., Salley, J.P., Hogquist, K.A., Chaplin, D.D., Mumford, R.A., Schmidt, J.A., and Tocci, M.J., 1991, IL-1-converting enzyme requires aspartic acid residues for processing of the IL-1β precursor at two distinct sites and does not cleave 31-kDa IL-1α, *J. Immunol.*, **147**:2964–2969.

Kamens, J., Paskind, M., Hugunin, M., Talanian, R.V., Allen, H., Banach, D., Bump, N., Hackett, M., Johnston, C.G., Li, P., Mankovich, J.A., Terranova, M., and Ghayur, T., 1995, Identification and characterization of ICH-2, a novel member of the interleukin-1β-converting enzyme family of cysteine proteases. *J. Biol. Chem.*, **270**:15250–15256.

Kaufmann, S.H., 1989, Induction of endonucleolytic DNA cleavage in human acute myelogenous leukemia cells by etoposide, camptothecin, and other cytotoxic anticancer drugs: A cautionary note, *Cancer Res.*, **49**:5870–5878.

Kaufmann, S.H., S. Desnoyers, Y. Ottaviano, N.E. Davidson and Poirier, G.G., 1993, Specific proteolytic cleavage of poly(ADP-ribose) polymerase: An early marker of chemotherapy-induced apoptosis, *Cancer Res.*, **53**:3976–3985.

Kerr, J.F.A., Wyllie, A.M., and Currie, A.R., 1972, Apoptosis: A basic phenomenon with wider ranging implications in tissue kinetics, *Br. J. Cancer*, **24**:239–275.

Kuida, K., Lippke, J.A., Ku, G., Harding, M.W., Livingston, D.J., Su, M.S.-S., Flavell, R.A., 1995, Altered cytokine export and apoptosis in mice deficient in interleukin-1β converting enzyme, *Science*, **267**:2000–2003.

Kumar, S., Kinoshita, M., Noda, M., Copeland, N.G. and Jenkins, N.A., 1994, Induction of apoptosis by the mouse *Nedd2* gene, which encodes a protein similar to the product of the *Caenorhabditis elegans* cell death gene *ced-3* and the mammalian IL-1β-converting enzyme, *Genes Dev.*, **8**:1613–1626.

Kwo, P., Patel, T., Bronk, S.F. and Gores, G.J., 1995, Nuclear serine protease activity contributes to bile acid-induced apoptosis in hepatocytes, *American Journal of Physiology Gastrointestinal & Liver Physiology*, **31**:

Lautier, D., Lagueux, J., Thibodeau, J., Ménard, L. and Porier, G.G., 1993, Molecular and biochemical features of poly(ADP-ribose) metabolism, *Mol. Cell. Biochem.*, **122**:171–193.

Lazebnik, Y.A., Cole, S., Cooke, C.A., Nelson, W.G., and Earnshaw, W.C., 1993, Nuclear events of apoptosis *in vitro* in cell-free mitotic extracts: a model system for analysis of the active phase of apoptosis, *J. Cell. Biol.*, **123**:7–22.

Lazebnik, Y.A., S.H. Kaufmann, Desnoyers, S., Poirier, G.G. and Earnshaw, W.C., 1994, Cleavage of poly(ADP-ribose) polymerase by a protease with properties like ICE, *Nature*, **371**:346–347.

Lazebnik, Y.A., Takahashi, A., Moir, R., Goldman, R., Poirier, G.G., Kaufmann, S.H., and Earnshaw, W.C., 1995, Studies of the lamin proteinase reveal multiple parallel biochemical pathways during apoptotic execution, *Proc. Natl. Acad. Sci. (USA)*, **92**:9042–9046.

Li, P., Allen, H., Banerjee, S., Franklin, S., Herzog, L., Johnston, C., McDowell, J., Paskind, M., Rodman, L., Salfeld, J., Towne, E., Tracey, D., Wardwell, S., Wei, F.-Y., Wong, W., Kamen R. and Seshadri, T., 1995, Mice deficient in IL-1β-converting enzyme are defective in production of mature IL-1β and resistant to endotoxic shock, *Cell*, **80**:401–411.

Lindahl, T., Satoh, M.S., Poirier, G.G. and Klungland, A., 1995, Post-translational modification of poly(ADP-ribose) polymerase induced by DNA strand breaks, *Trends Biochem. Sci.*, **20**:405–411.

Ludérus, M.E.E., de Graaf, A., Mattia, E., den Blauuwein, J.L., Grande, M.A., de Jong, L. and van Driel, R., 1992, Binding of matrix attachment regions to lamin B_1, *Cell*, **70**:949–959.

Martin, S.J., Mazdai, G., Strain, J.J., Cotter, T.G., and Hannigan, B.M., 1991, Programmed cell death (apoptosis) in lymphoid and myeloid cell lines during zinc deficiency, *Clin Exp Immuno*, **83**:338–43.

McCabe, M.J., Jiang, S.A., and Orrienius, S., 1993, Chelation of intracellular zinc triggers apoptosis in mature thymocytes, *Lab Invest*, **69**:101–10.

Miura, M., Zhu, H., Rotello, R., Hartweig, E.A., and Yuan, J., 1993, Induction of apoptosis in fibroblasts by IL-1β-converting enzyme, a mammalian homologue of the C. elegans cell death gene *ced-3*, *Cell*, **75**:653–660.

Moir, R.D., Donaldson, A.D. and Steward, M., 1991, Expression in Escherichia coli of human lamins A and C: influence of head and tail domains on assembly properties and paracrystal formation, *J Cell Sci*, **99**:363–372.

Munday, N.A., Vaillancourt, J.P., Ali, A., Casano, F.J., Miller, D.K., Molineaux, S.M., Yamin, T.-T., Yu, V.L. and Nicholson, D.W., 1995, Molecular cloning and pro-apoptotic activity of ICE$_{rel}$II and ICE$_{rel}$III, members of the ICE/CED-3 family of cysteine proteases, *J. Biol. Chem.*, **270**: 15870–15876.

Newmeyer, D.D., D.M. Farschon and J.C. Reed, 1994, Cell-free apoptosis in Xenopus egg extracts – inhibition by Bcl-2 and requirement for an organelle fraction enriched in mitochondria, *Cell*, **79**:353–364.

Nicholson, D.W., Ali, A., Thornberry, N.A., Vaillancourt, J.P., Ding, C.K., Gallant, M., Gareau, Y., Griffin, P.R., Labelle, M., Lazebnik, Y.A., Munday, N.A., Raju, S.M., Smulson, M.E., Yamin, T.-T., Yu, V.L. and Miller, D.K., 1995, Identification and inhibition of the ICE/CED-3 protease necessary for mammalian apoptosis, *Nature*, **376**:37–43.

Ozaki, T., Saijo, M., Murakami, K., Enomoto, H., Taya, Y. and Sakiyama, S., 1994, Complex formation between lamin α and the retinoblastoma gene product – identification of the domain on lamin α required for its interaction, *Oncogene*, **9**:2649–2653.

Pattison, S.E., and Cousins, R.J., 1986, Zinc uptake and metabolism by hepatocytes, *Fed. Proc.*, **45**:2805–2809.

Ray, C.A., Black, R.A., Kronheim, S.R., Greenstreet, T.A., Sleath, P.R., Salvesen, G.S., and Pickup, D.J., 1992, Viral inhibition of inflammation: Cowpox virus encodes an inhibitor of the interleukin-1β converting enzyme, *Cell*, **69**:597–604.

Satoh, M., Poirier, G.G. and Lindahl, T., 1994, Dual function for poly(ADP-ribose) synthesis in response to DNA strand breakage, *Biochem.*, **33**:7099–7106.

Sleath, P.R., Hendrickson, R.C., Kronheim, S.R., March, C.J., and Black, R.A., 1990, Substrate specificity of the protease that processes human interleukin-1 beta, *J. Biol. Chem.*, **265**:14526–14528.

Sun, X.M., Snowden, R.T., Dinsdale, D., Ormerod, M.G., and Cohen, G.M., 1994, Changes in nuclear chromatin precede internucleosomal DNA cleavage in the induction of apoptosis by eptoposide, *Biochem Pharmacol*, **47**:187–95.

Sunderman, F.W., 1995, The influence of zinc on apoptosis, *Annals of Clinical & Laboratory Science*, **25**:134–142.

Takahashi, A., Alnemri, E., Lazebnik, Y.A., Fernandes-Alnemri, T., Litwack, G., Moir, R.D., Goldman, R.D., Poirier, GG., Kaufmann, S.H. and Earnshaw, W.C., 1996, Activity of multiple ICE-related proteases with distinct substrate recognition properties in apoptosis: Cleavage of lamins by Mch2α but not CPP32. *Proc. Natl. Acad. Sci. (USA)*, **93**:8395–8400.

Takano, Y.S., Harmon, B.V. and Kerr, J.F., 1991, Apoptosis induced by mild hyperthemia in human and murine tumour cell lines: a study using electron microscopy and DNA gel electrophoresis, *J. Pathol.*, **163**:329–36.

Taniura, H., Glass, C. and Gerace, L., 1995, A chromatin binding site in the tail domain of nuclear lamins that interacts with core histones, *J. Cell. Biol.*, **131**:33–44.

Tewari, M., Beilder, D.R., and Dixit, V.M., 1995, CrmA-inhibitable cleavage of the 70-kDa protein component of the U1 small nuclear ribonucleoprotein during Fas- and tumor necrosis factor-induced apoptosis, *J. Biol. Chem.*, **270**:18738–18741.

Tewari, M., Quan, L.T., O'Rourke, K., Desnoyers, S., Zeng, Z., Beilder, D.R., Poirier, G.G., Salvesen, G.S. and Dixit, V.M., 1995, Yama/CPP32β, a mammalian homolog of CED-3, is a CrmA-inhibitable protease that cleaves the death substrate poly(ADP-ribose) polymerase, *Cell*, **81**:801–809.

Thornberry, N.A., Bull, H.G., Calycay, J.R., Chapman, K.T., Howard, A.D., Kostura, M.J., Miller, D.K., Molineaux, S.M., Weidner, J.R., Aunins, J., Elliston, K.O., Ayala, J.M., Casano, F.J., Chin, J., Ding, G.J.-F., Egger, L.A., Gaffney, E.P., Limjuco, G., Palyha, O.C., Raju, S.M., Rolando, A.M., Salley, J.P., Yamin, T.-T., Lee, T.D., Shively, J.E., MacCross, M., Mumford, R.A., Schmidt, J.A., and Tocci, M.J., 1992, A novel heterodimeric cysteine protease is required for interleukin-1β processing in monocytes. *Nature*, **356**:768–774.

Thornberry, N.A., Peterson, E.P., Zhao, J.J., Howard, A.D., Griffin, P.R., and Chapman, K.T., 1994, Inactivation of interleukin-1β converting enzyme by peptide (acyloxyl)methyl ketones, *Biochemistry*, **33**:3934–3940.

Treves, S., Trentini, P.L., Ascanelli, M., Bucci, G. and Di, G.F., 1994, Apoptosis is dependent on intracellular zinc and independent of intracellular calcium in lymphocytes, *Exp Cell Res*, **211**:339–43.

Ucker, D.S., Meyers, J. and Obermiller, P.S., 1992, Activation driven T cell death. II. Quantitative differences alone distinguish stimuli triggering nontransformed T cell proliferation or death, *J. Immunol.*, **149**:1583–1592.

Walker, N.P.C., Talanian, R.V., Brady, K.D., Dang, L.C., Bump, N.J., Ferenz, C.R., Franklin, S., Ghayur, T., Hackett, M.C., Hammill, L.D., Herzog, L., Hugunin, M., Houy, W., Mankovich, J.A., McGuiness, L., Orlewicz, E., Paskind, M., Pratt, C.A., Reis, P., Summani, A., Terranova, M., Welch, J.P., Xiong, L., Möller, Tracey, D.E., Kamen, R. and Wong, W.W., 1994, Crystal structure of the cysteine protease interleukin-IL-1β converting enzyme: A (p20/p10)$_2$ homodimer, *Cell*, **78**:343–352.

Wang, L., Miura, M., Bergron, L., Zhu, H. and Yuan, J., 1994,Ich-1 an Ice/ced-3-related gene encodes both positive and negative regulators of programmed cell death, *Cell*, **78**:739–750.

Wang, X., Pai, T.-T., W. E.A., Medina, J.C., S.C.A., Goldstein, J.L. and Brown, M.S., 1995, Purification of an interleukin-1β converting enzyme-related cysteine protease that cleaves sterol regulatory element-binding proteins between the leucine zipper and transmembrane domains, *J. Biol. Chem.*, **270**:18044–18050.

Waring, P., Egan, M., Braithwaite, A., Mullbacher, A. and Sjaarda, A., 1990, Apoptosis induced in macrophages and T blasts by the mycotoxin sporidesmin and protection by Zn^{2+} salts, *Int. J. Immunophamac.*, **12**:445–457.

Williams, M.S., and Henkart, P.A., 1994, Apoptotic cell death induced by intracellular proteolysis, *J. Immunol.*, **153**:4247–4255.

Wilson, K.P., Black, J.-A., Thomson, J.A., Kim, E.E., Griffith, J.P., Navia, M.A., Murcko, M.A., Chambers, S.P., Aldape, R.A., Raybuck, S.A. and Livingston, D.J., 1994, Structure and mechanism of interleukin-1β converting enzyme, *Nature*, **370**:270–275.

Wood, E.R. and Earnshaw, W.C., 1990, Mitotic chromatin condensation *in vitro* using somatic cell extracts and nuclei with variable levels of endogenous topoisomerase II, *J. Cell Biol.*, **111**:2839–2850.

Wyllie, A.H., 1980, Glucocorticoid-induced thymocyte apoptosis is associated with endogenous endonuclease activation, *Nature*, **284**:555–556.

Wyllie, A.H., Kerr, J.F.R., and Currie, A.R., 1980, Cell death: The significance of apoptosis, *Int. Rev. Cytol.*, **68**:251–305.

Yamanaka, H., Penning, C.A., Willis, E.H., Wasson, D.B. and Carson, D.A., 1988, Characterization of human poly(ADP-ribose) polymerase with autoantibodies, *J. Biol. Chem.*, **263**:3879–3883.

Yuan, J.Y. and Horvitz, H.R., 1990, The Caenorhabditis elegans genes ced-3 and ced-4 act cell autonomously to cause programmed cell death, *Dev. Biol.*, **138**:33–41.

Yuan, J., Shaham, S., Ledoux, S., Ellis, H.M., and Horvitz, H.R., 1993, The C. elegans cell death gene *ced-3* encdes a protein similar to mammalian interleukin-1β-converting enzyme, *Cell*, **75**:641–652.

Zalewski, P.D., Forbes, I.J., and Giannakis, C., 1991, Physiological role for zinc in prevention of apoptosis (gene-directed death), *Biochem Int*, **24**:1093–101.

10

Bcl-2 Family Proteins in Cancer: Regulators of Cell Death Involved in Resistance to Therapy

JOHN C. REED

1. Introduction

Cell death is a physiological process that plays a critical role in the regulation of tissue homeostasis by ensuring that the rate at which new cells are produced in the body through cell division is offset by a commensurate rate of cell loss. The amount of cell death that occurs constantly within cell-renewing tissues such as a bone marrow, gut, and skin is enormous. In fact, some estimates suggest that in the course of a typical year, each of us will lose, through cell death, and have replenished, through cell division, a mass of cells equivalent to our entire body weight. Though largely overlooked until recently, it is now increasingly appreciated that disturbances in the physiological cell death process that prevent or delay normal cell turnover can be just as important to the pathogenesis of cancer as abnormalities in the regulation of the cell cycle. Perhaps of even greater clinical importance is the recent realization that not only are defects in the cell death pathway important in the origins of cancer, but they may also markedly influence our ability to treat it. Because nearly all chemotherapeutic drugs, as well as radiation, ultimately tap into endogenous physiological pathways for cell death to kill cancer cells, the loss of genes required for cell death or the over-activation of genes that block it can render tumor cells relatively more resistant to the cytotoxic effects of a broad spectrum of anti-cancer drugs.

Tel: 619-646-3140 (x430); Fax: 619-646-3194; Email: jreed@ljcrf.edu

2. The Discovery of BCL-2 in Human Lymphomas and the Delineation of a Family of Homologous Genes

Like cell division, which is controlled through a complex interplay of cell cycle stimulators and repressors, the physiological cell death pathway is precisely regulated under normal circumstances by a delicate balance of genes whose encoded proteins either induce or inhibit cell death. The first realization that defects in the cell death pathway could contribute to the development of human neoplasia came from the discovery of a gene involved frequently in non-Hodgkin's lymphomas, called *BCL-2* for "B-cell lymphoma-2" (Tsujimoto 1986; Vaux, 1988). In approximately 90% of low-grade follicular non-Hodgkin's lymphomas, as well as approximately 30% of more aggressive B-cell lymphomas, chromosomal translocations move the *BCL-2* gene from its normal location on chromosome 18 into juxtaposition with the immunoglobulin (Ig) heavy-chain gene locus on chromosome 14, probably as the result of errors in the normal DNA recombination mechanisms that cut and splice together the V, D., and J gene segments to create functional Ig genes during B-cell differentiation in the bone marrow (Tsujimoto *et al.*, 1988). The resulting t(14;18) translocations place the *BCL-2* gene under the influence of powerful transcriptional enhancers associated with the Ig locus, thus dysregulating the expression of *BCL-2*, primarily at the transcriptional level. Because the protein encoded by the *BCL-2* gene blocks programmed cell death, B-cells containing a t(14;18) translocation enjoy a selective survival advantage relative to their normal counterparts, and begin to clonally expand without necessarily experiencing an increase in their doubling times. When one considers that the average life-span of B-cells is only 5 to 7 days, it becomes immediately obvious how a genetic alteration that prevents cell death can impact on the homeostatic mechanisms that control the number of these cells in the body. This mechanism for clonal expansion, which is based on a selective survival advantage as opposed to an increased rate of cell division, probably explains the low growth fraction of most follicular lymphomas, which are generally regarded as low-grade malignancies in which the accumulation of malignant B-cells in the body occurs slowly over time but nevertheless ultimately leads to patient demise.

The protein encoded by the *BCL-2* gene is unique, and has no significant amino-acid homology with other proteins whose biochemical mechanism of action is known. Comparisons of the sequences of the human, mouse, rat, and chicken homologs of Bcl-2 have suggested a four-domain structure where an approximately 40 amino acid conserved N-terminal domain is followed by a non-conserved region of variable length that is often rich in prolines and thus unlikely to fold into higher-order structures such as α-helies or β-sheets. This is followed by another well-conserved region of approximately 100 amino-acids length, and then a stretch of hydrophobic amino acids at the C-terminus that has been shown to constitute a transmembrane domain (Cazals-Hatem *et al.*, 1992). Thus, Bcl-2 is an integral membrane protein. The intracellular membranes into which the Bcl-2

Table I. Bcl-2 Family Proteins.

Protein	Source	Primary Function
Bcl-2	mammalian	anti-apoptotic
Bcl-X$_L$	mammalian	anti-apoptotic
Bcl-X$_S$	mammalian	pro-apoptotic
Bax	mammalian	pro-apoptotic
Mcl-1	mammalian	anti-apoptotic
A1	mammalian	anti-apoptotic
Bak	mammalian	pro-apoptotic
Bik	mammalian	pro-apoptotic
Nr13	avian	anti-apoptotic
ced9	nematode	anti-apoptotic
E1b-19kDa	adenovirus	anti-apoptotic
BHRF1	Epstein Barr Virus	anti-apoptotic
LMW5-HL	African Swine Fever Virus	not tested
ORF-16	Herpes virus saimiri	not tested

protein post-translationally inserts include predominantly the outer mitochondrial membrane, nuclear envelope, and parts of the endoplasmic reticulum (Krajewski *et al.*, 1993). Although mutagenesis studies suggest that the transmembrane (TM) domain of Bcl-2 is essential for optimal function in other types of cells, under some circumstances C-terminal truncation mutants of Bcl-2 that lack a membrane anchor are as effective as the wild-type Bcl-2 proteins at blocking cell death (Tanaka *et al.*, 1993; Borner *et al.*, 1994). However, such TM-deficient versions of Bcl-2, probably still retain the ability to interact with other membrane-associated proteins in cells, and thus may find their way to appropriate intracellular locations through protein-protein interactions.

Since the discovery of *BCL-2*, several homologs of this gene and its encoded protein have been identified (Table I). Interestingly, many of these Bcl-2-related proteins can physically interact with each other in the form of homo- and heterotypic dimers or oligomers (the actual stoichiometry is unknown at present). Furthermore, some of these homologs function as blockers of cell death, whereas others are promoters of apoptosis. At present, eight mammalian homologs of Bcl-2 have been reported: Bax, Nr13, Bik, Bcl-X, Mcl-1, A1, Bad, and Bak (Oltvai *et al.*, 1993; Boise *et al.*, 1993; Kozopas *et al.*, 1993; Lin *et al.*, 1993, Yang *et al.*, 1995, Chittenden *et al.*, 1995; Kiefer *et al.*, 1995; Farrow *et al.*, 1995; Gillet *et al.*, 1995). Some of these proteins have additional forms that arise through alternative splicing, the most interesting of which are the long (L) and short (S) forms of Bcl-X. The Bcl-X$_L$ protein (which is 47% identical to Bcl-2 at the amino acid level) and the Bcl-X$_S$ protein (which is missing a well-conserved 63 amino acid region) have opposing functions, with Bcl-X$_L$ acting as a cell death blocker and Bcl-X$_S$ as an antagonist of Bcl-2 and Bcl-X$_L$ that promotes apoptosis (Boise *et al.*, 1993). In addition to Bcl-X$_S$, the Bax, Bad, Bik, and Bak proteins function as promoters of cell death. Conversely, the Mcl-1 and A1 proteins appear to be suppressors of cell death, like Bcl-2 and Bcl-X$_L$. Finally, several homologs of Bcl-2 have been discovered in viruses, including the E1b-19kD protein of adenovirus and the BHRF-1 protein of Epstein Barr Virus (EBV) – both of which function as suppressors of cell death (Chiou et al., 1994; Takayama et al., 1994). The sparse economy of viral genomes implies that intense evolutionary pressures must have selected for retention of these viral homologs of Bcl-2, and suggests that Bcl-2 represents a critical point for regulating the physiological cell death pathway.

When expressed in yeast (Saccharomyces cerevisiae), the Bax protein confers a lethal phenotype, suggesting that it promotes cell death through a mechanism that may be evolutionarily conserved (Sato et al., 1994; 1995). The Bak protein also has this effect in yeast, whereas Bcl-x$_S$ does not. Co-expression in yeast of fusion proteins that represent Bcl-2, Bcl-x$_L$ or Mcl-1 without their TM domains neutralizes Bax-mediated cytotoxicity; thus, a TM domain is not absolutely required for any of these Bcl-2 family proteins in this system (Sato et al., 1994; 1995; Hanada et al., 1995; Bodrug et al., 1995). Conversely, the TM domain of Bax is required for lethality in yeast. Mutagenesis studies have shown that at least three conserved domains in Bcl-2 are required for its ability to suppress Bax-mediated death in yeast (Hanada et al., 1995). These three domains are generally well conserved among anti-apoptotic members of the Bcl-2 protein family, and are termed BH1, BH2, and BH4 in Fig. 1. Investigations of equivalent deletion mutants of Bcl-2 in mammalian cells have produced similar results, showing an inability to protect cells from apoptotic stimuli (Borner et al., 1994). Interestingly, co-immunoprecipitation and other experiments have demonstrated that some of these Bcl-2 deletion mutants, such as those lacking BH1 and BH2, have lost the ability to heterodimerize with Bax (Yin et al., 1994; Hanada et al., 1995). Thus, binding to Bax appears to be one important feature of Bcl-2 function. However, in vitro binding studies indicate that some Bcl-2 deletion mutants, such as those lacking the BH4 domain, still bind to Bax and appear to do so with roughly the same efficiency as the wild-type Bcl-2 protein, although quantitative measurement of affinities have not been performed (Hanada et al., 1995). Consequently, it appears that while binding to Bax may be important for Bcl-2 function, it is not the only requirement. What those other requirements are remains to be determined, but they may include binding to other proteins by Bcl-2 or masking sites on Bax so that other proteins cannot bind to or post-translationally modify Bax. In this regard, the BH4 domain of Bcl-2 appears to be required for interactions with a novel Bcl-2 binding protein, BAG-1, which functionally cooperates with Bcl-2 in co-transfection assays, enhancing Bcl-2's ability to block cell death (Takayama et al., 1995; and unpublished data).

Though Bax appears to directly promote cell death though still poorly understood mechanisms, another class of Bcl-2 homologs, which includes Bcl-X$_S$ and Bad, indirectly induces apoptosis by binding to Bcl-2 and Bcl-X$_L$ and preventing them from heterodimerizing with Bax (see Fig. 2). The complexity

STRUCTURES OF Bcl-2 FAMILY PROTEINS

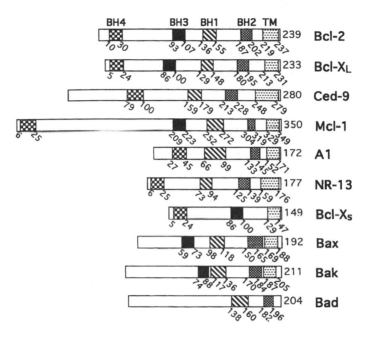

Figure 1. Functional domains in Bcl-2 family proteins. The known cellular members of the Bcl-2 protein family are depicted, showing the relative locations of the BH1, BH2, BH3, BH4 and transmembrane (TM) domains.

of these interactions among Bcl-2 family proteins undoubtedly have evolved to provide multiple opportunities for fine-tuning the relative sensitivity or resistance of cells to apoptotic stimuli through differential regulation of the expression of various *BCL-2* family genes (Sato *et al.*, 1994; Yang *et al.*, 1995).

3. Bcl-2 and Chemoresistance

We have used gene transfer methods to over-express *BCL-2* in leukemic and solid tumor cell lines that normally contain low levels of Bcl-2 protein, as well as antisense approaches to reduce the levels of Bcl-2 protein in t(14;18)-containing lymphoma cell lines that contain high levels of this protein, and we have shown that levels of Bcl-2 protein correlate with relative resistance to a wide spectrum of chemotherapeutic drugs as well as γ-irradition (Miyashita *et al.*, 1992; 1993; Hanada *et al.*, 1993; Kitada *et al.*, 1994). Bcl-2 has been experimentally shown

**Model for Effects of Heterodimerization of
Bcl-2 Family Proteins on Cell Death Regulation.**

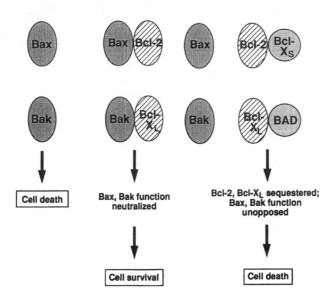

Figure 2. Model for Bcl-2 family protein interactions. Based on evidence available to date, a model can be envisioned in which Bax promotes apoptosis. Though shown here as a monomer, Bax possibly must homodimerize to function. Bax-mediated cell death is opposed when Bcl-2, Bcl-X_L, Mcl-1, A1 or possibly other homologs of Bcl-2 that have anti-death activity heterodimerize with this pro-apoptotic protein, thus neutralizing its function. A second class of cell death promoters, which include Bcl-X_S and Bad, indirectly induce apoptosis by binding to Bcl-2, Bcl-X_L and probably other anti-apoptotic members of the Bcl-2 protein family, thus sequestering them and preventing them from heterodimerizing with Bax. This leaves Bax homodimers unopposed. A recently described member of the Bcl-2 protein family, Bak, may function equivalent to Bax, but appears to bind with higher affinity to Bcl-x_L than Bcl-2. The Bad protein also has higher affinity for Bcl-x_L than Bcl-2.

to render cells more resistant to the following drugs: dexamethasone, cytosine arabinoside (Ara-C), methotrexate, cyclophosphamide, adriamycin, daunomycin, 5-fluoro-uracil, 2-chlorodeoxyadenosine, fludarabine, taxol, etoposide (VP-16), camptothecin, nitrogen mustards, mitoxantrone, and cisplatin. The extent to which Bcl-2 provides protection from the cytotoxic effects of these drugs varies, depending on the particular drug and the cell line, but it can be as much as four or more logs (10,000x), or as little as half a log (5x). Given that so-called "high-dose" aggressive chemotherapy typically involves a mere doubling of the concentration of drugs delivered to patients, even a five-fold increase in relative resistance could be enormously significant in clinical terms.

Table II. Mechanisms of Chemoresistance.

---

1. Increased drug efflux (Mdr-1).
2. Metabolism to inactive products; failure to metabolize pro-drug to active compound.
3. Alteration of levels or structure of primary target (dhfr gene amplication; estrogen receptor loss or mutation).
4. Increased repair of drug-induced damage (DNA repair enzymes).
5. Decreased primary damage (glutathione).
6. Decreased apoptotic response to drug-induced damage (Bcl-2).

The observation that Bcl-2 provides protection against such a wide variety of drugs with markedly diverse mechanisms of action suggests that they all utilize the same final common pathway for ultimately inducing cell death, and that Bcl-2 is a regulator of this pathway. Indeed, several studies have provided evidence that when chemotherapeutic drugs, as well as γ-radiation, are administered *in vitro* to tumor cell lines, they induce cell death through mechanisms consistent with apoptosis as opposed to necrosis. It stands to reason, therefore, that genes such as *BCL-2*, which block the apoptotic pathway, could also block cell killing induced by anti-cancer drugs.

The mechanism by which Bcl-2 confers resistance to anticancer drugs is distinct from other previously recognized forms of chemoresistance. Traditionally, pharmacologists have thought of the chemoresistance problem in cancer in terms of four issues: (1) delivery of drug to the target, such as occurs when the *mdr-1* gene product, P-glycoprotein, is over-produced in the plasma membrane of cancer cells and pumps drugs out of the cell or when a drug is metabolized to an inactive product; (2) modification of the drug target, an example of which is amplification of the gene for dihydrofolate reductase, which often occurs following exposure to methotrexate; (3) increased rates of repair of damage to DNA or other structures; and (4) diminished rates of drug-induced damage to DNA or other macromolecules, as can occur for some drugs when glutathione levels are elevated in tumors. (Table II). Bcl-2, in contrast, does not appear to interfere with the ability of drugs of enter cells, bind to their appropriate targets, and induce damage. Indeed, Bcl-2 does not protect cancer cells from drug-induced cell cycle arrest but does prolong their survival during this period, so that proliferation can resume upon withdrawal of the drug, as typically occurs between cycles of chemotherapy. Rates of DNA repair are also not affected by Bcl-2. In cells that over-produce Bcl-2, therefore, anticancer drugs induce cell cycle arrest and damage to DNA or other macromolecules, but this damage somehow is not translated effectively into signals for cell death. As such, Bcl-2 defines a new category of chemoresistance gene; namely, those that regulate downstream events in the normal physiological pathway for programmed cell death and that convert anticancer drugs from cytotoxic to merely cytostatic. The extent to which Bcl-2 controls treatment outcomes, remains to be determined within the clinical context of patients, but at least some clinical correlative studies have suggested that there is a connection between Bcl-2 and either poor response to therapy, shorter disease-free survival, or shorter overall survival in some groups of patients with large cell non-Hodgkin's lymphomas, myeloid leukemias, and adenocarcinomas of the prostate (reviewed in Reed *et al.*, 1994, 1995).

4. Mechanisms of BCL-2 Gene Dysregulation in Human Cancers

Although the *BCL-2* gene was first discovered because of its involvement in t(14;18) translocations found frequently in non-Hodgkin's lymphomas, high levels and aberrant patterns of *BCL-2* gene expression have been reported in a wide variety of human cancers, including approximately 90% of colorectal, 60% of gastric, 30%-60% of prostate, 20% of non-small cell lung cancers, 30% of neuroblastomas, and variable percentages of melanomas, renal cell, and thyroid cancers, as well as acute and chronic lymphocytic and non-lymphocytic leukemias (reviewed in Reed *et al.*, 1994, 1995). In essentially all of these non-lymphomatous cancers, no evidence for structural alterations of the *BCL-2* gene has been found; instead, alterations in trans-acting factors that control *BCL-2* gene expression are suspected to be at fault.

One of the mechanisms that may play a role in the dysregulation of *BCL-2* expression in cancers is the loss of the tumor suppressor p53. Loss of p53 function occurs in about half of all human cancers. This DNA-binding protein can both induce cell cycle arrest and promote apoptosis, and functions at least in part as a transcriptional regulator. In experiments in which p53 function was conditionally restored to a p53-deficient murine leukemia line, p53 was shown to induce marked decreases in *bcl-2* gene expression followed by apoptotic cell death (Miyashita *et al.*, 1994a). When Bcl-2 protein levels were maintained at high levels through gene transfer manipulations, p53 induced apoptosis was partially blocked but cell cycle arrest occurred normally (Selvakumaran, 1994). γ-radiation, a known inducer of p53, has also been shown to downregulate *BCL-2* mRNA levels in a human leukemia line (Zhan *et al.*, 1994). Thus, p53 either directly or indirectly appears to be able to suppress *BCL-2* gene expression, leading to the speculation that p53 loss in human tumors may contribute to the high levels and abnormal patterns of Bcl-2 protein production observed in many types of cancer. Indeed, using reporter gene assays, we have mapped a p53-negative response element (NRE) to the 5'-untranslated region of the *BCL-2* gene (Miyashita *et al.*, 1994b). However, our analysis of p53 knock-out mice suggests that the loss of p53 may be sufficient to result in elevated levels of Bcl-2 protein production in only some tissues, suggesting either that other p53-independent mechanisms for repression of *BCL-2* gene expression exist or that critical transactivators of *BCL-2* are missing from some types of cells (Miyashita *et al.*, 1994a). (Fig. 3).

Model for Promotion of Apoptosis by p53

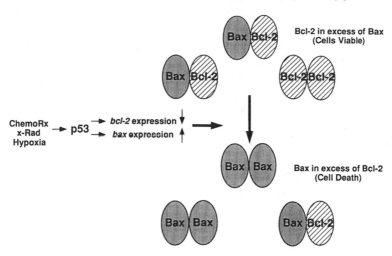

Figure 3. Model for regulation of chemosensitivity by tumor suppressor p53. Radiation, DNA-damaging chemotherapeutic drugs, and hypoxia are known to induce increases in p53 protein and p53 transcriptional activity. These elevations in p53 then directly upregulate *BAX* gene transcription and also downregulate *BCL-2* gene expression, at least in some types of cells and tissues where the effects of p53 are dominant to other factors that operate at the transcriptional or post-transcriptional levels to modulate Bax protein levels. As a consequence, the ratio of Bcl-2 to Bax protein declines, resulting in an excess of Bax protein and promotion of apoptotic cell death.

5. Dysregulation of Expression of Other BCL-2 Family Genes in Cancer

In addition to inhibiting *BCL-2* gene expression, the tumor suppressor p53 can also induce increases in *BAX* gene expression (Miyashita *et al.*, 1994a). These effects of p53 on *BCL-2* and *BAX* gene expression can result in a marked decrease in the ratio of Bcl-2 to Bax protein, and thus render cells more vulnerable to apoptotic stimuli (see Fig. 3). The *BAX* gene promoter contains four 10 bp motifs with homology the consensus p53-binding sites, and is strongly transactivated by p53 in reporter gene assays (Miyashita and Reed, 1995). Thus, *BAX* represents the first co-apoptotic gene to be identified that is a direct transcriptional target of p53 (Fig. 3). Clearly, however, p53 represents only one of the inputs into the *BAX* gene promoter, and other underlineated factors may modulate the effects of p53 on this gene. In this regard, radiation has been shown to induce expression of genes associated with gentoxic stress and cell cycle arrest in a p53-dependent fashion in many types of tumor cell lines, but triggers elevations in *BAX* mRNA and apoptosis only in a subset of cancer lines *in vitro* (Zhan, 1994). Similarly, γ-radiation in mice induces marked elevations in Bax protein levels only in radiosensitive tissues such as lymphoid organs and small intestine, but not in radioresistant tissues such as liver, kidney, and muscle (Kitada *et al.*, 1995). In contrast, γ-radition does stimulate increases in p53 target genes such as p21-WAF1 in a wide variety of tissues, both radioresistant and radiosensitive (Macleod *et al.*, 1995). The mechanisms that prevent p53 from transactivating the *BAX* gene in many tumor lines and tissue remains to be determined, but once delineated it could provide insights into strategies for improving tumor responses to radiotherapy and DNA-damaging chemotherapeutic drugs. Interestingly, hypoxia has also been reported to induce increases in p53 in the brain and in a model of transient global ischemia in the rat, and Bax immunostaining was found to be strikingly elevated in dying neurons (Krajewski *et al.*, 1995b). It was not determined, however, whether p53 mediated this upregulation of Bax protein in ischemic neurons.

It seems likely that since p53 can bind to and transactivate the *BAX* gene, tumors with loss of p53 function will contain relatively lower levels of Bax protein. Indeed, we have observed that Bax protein levels are markedly reduced in about one-third of advanced breast cancers, and that these reductions in Bax correlate with poor responses to chemotherapy and shorter overall survival, at least in some subgroups of patients (Krajewski *et al.*, 1995a). However, reduced levels of Bax expression did not correlate with p53-mutations, suggesting that other mechanisms may be responsible. In contrast, Bax reductions do correlated with decreases in Bcl-2 expression, suggesting that these proteins may be somewhat co-regulated (Miyashita *et al.*, 1995), and possibly providing an explanation for the paradoxical association of decreased Bcl-2 protein levels with poor clinical outcomes for patients with breast cancer (Silvestrini *et al.*, 1994). In this regard, we have observed that tumors with diminished Bax and Bcl-2 typically express Bcl-X_L and Mcl-1. Thus, despite the decrease in Bcl-2, the ratio of anti-apoptotic Bcl-2 family proteins to pro-apoptotic proteins is probably increased in advanced breast cancers, thus accounting for the poorer responses of these patients to therapy and their shorter overall survival (Krajewski *et al.*, 1995) (Fig. 4). Finally, we have observed prominent increases in Bcl-X_L protein production (rather than elevations in Bcl-2 or decreases in Bax), in association with drug-resistant phenotypes in some leukemias and solid tumors, implying that alterations in the expression of Bcl-X_L may also be relevant to mechanisms of drug resistance in other types of cancer. Although these studies are just beginning, the preliminary observations suggest that dysregulation of the expression of several members of the *BCL-2* gene family is likely to contribute directly or indirectly to drug resistance in cancers.

6. Possible Mechanisms of Bcl-2 Protein Action

At present, a biochemical understanding of how Bcl-2 and its homologs control cell life and death remains elusive (Table III). Indirect evidence suggests that Bcl-2 has an effect on the regulation of antioxidant pathways in cells and particularly protects

Table III. Potential Mechanisms
of Bcl-2 Protein Action.

---

1. Antioxidant pathway
2. Ca^{2+} transport
3. Protein translocation
4. Proteases
5. Signal transduction
6. None of the above

Bcl-2 Family Proteins in Breast Cancer

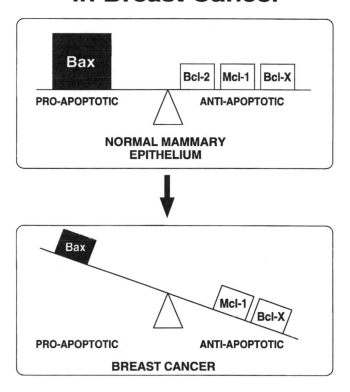

Figure 4. Imbalance in Bcl-2 family proteins occurs during progression of breast cancers. In normal mammary epithelium, cell death pathways are appropriately regulated by a balance between pro-apoptotic Bcl-2 family proteins such as Bax and anti-apoptotic proteins including Bcl-2, Bcl-X_L and Mcl-1. During progression of breast cancers, Bax and Bcl-2 protein levels become reduced but Bcl-X_L and Mcl-1 remain. The net effect is a tilting of the balance in favor of the anti-cell death proteins.

against lipid peroxidation (Kane, 1993; Hockenbery, 1993). Recent observations that Bcl-2 protects against apoptosis even under conditions of hypoxia has reduced enthusiasm for this hypothesis, but this does not exclude the possibility of thiol-based redox mechanisms (Jacobson and Raff, 1995; Shimizu et al., 1995). Experimental evidence suggesting the regulation of intracellular Ca^{2+} homeostasis has also been obtained. In a hemopoietic cell in which apoptosis is induced by lymphokine deprivation and in glucocorticoid-treated lymphoid cells, massive loss of Ca^{2+} from the ER occurs as a relatively early event prior to apoptosis. Enforced production of high levels of Bcl-2 protein in these cells delays apoptosis and loss of ER Ca^{2+} stores (Baffy, 1993; Lam, 1994). In the hemopoietic cell model, Bcl-2 was also found to influence mitochondrial pools of Ca^{2+}, preventing an accumulation of Ca^{2+} in this organelle that can serve as a sink for Ca^{2+} under conditions of highly cytosolic Ca^{2+} concentrations (Baffy, 1993). Furthermore, Bcl-2 was reported to delay the efflux of Ca^{2+} from the ER in cells treated with thapsigargin, a specific inhibitor of that organelle's Ca^{2+}-ATPase. Although no direct link between Bcl-2 and Ca^{2+}-channels or other Ca^{2+}-regulating proteins has been found, it is interesting that two recently described Bcl-2-binding proteins, Nip-2 and Nip-3, contain sequences that resemble Ca^{2+}-binding sites or that have homology to calbindin-D (Boyd, 1994), an ER-protein that has been shown to delay apoptosis when over-expressed in glucocorticoid-sensitive lymphoid cells (Dowd, 1992). Of course, it is possible that the effects of Bcl-2 on lipid peroxidation and Ca^{2+} transport are related, since Ca^{2+} can influence the activity of some enzymes involved in lipid metabolism, and oxidative damage to membranes can compromise Ca^{2+} compartmentalization.

It has also been suggested that Bcl-2 may participate in protein transport across biological membranes (Reed, 1994). For example, immunoelectromicroscopic studies indicate that Bcl-2 is located in discrete patches distributed non-uniformly in the outer mitochondrial membrane and nuclear envelope. This is not unlike proteins targeted to the mitochondrial junctional complexes (MJCs) and nuclear pore complexes (NPCs), where the inner and outer membranes of these DNA-containing organelles come into contact and where transport of peptides, RNA, and probably some ions occurs (Krajewski, 1993; deJong et al., 1994). In this regard, the nuclear accumulation of p53 and some cyclin-dependent kinases has been reported to be antagonized by gene transfer-mediated elevations in Bcl-2 protein (Ryan et al., 1994; Meikrantz et al., 1994), but several, other groups have failed to find any effects of Bcl-2 on the translocation of temperature-sensitive versions of p53 from cytosol to nucleus. Studies using enucleated cells, as well as a cell-free system involving apoptotic cytosolic extracts from xenopus eggs, have also provided convincing support for the idea that apoptosis is largely a cytoplasmically regulated process with the nucleus serving as a mere passive substrate for degradation, at least in circumstances under which the induction of cell death does not require new gene expression (Jacobson et al., 1994a; Newmeyer et al., 1994).

A link between Bcl-2 and the regulation of protease has also been suggested both by genetic studies in the nematode *C. elegans* and gene transfer studies in mammalian cells, where cell death induced by *ced-3*, a cysteine protease, and its homologs was shown to be inhibitable by either Bcl-2 or its equivalent in the worm *ced-9* (Yuan *et al.*, 1993; Miura *et al.*, 1993). Bcl-2 has also been shown to block proteolytic processing of at least some ICE/ced-3 family proteases from their inactive zymogens to active proteases, further strengthening arguments that Bcl-2 may function via effects on these enzymes. The discovery of a mammalian Bcl-2-binding protein, BAG-1, which contains a ubiquitin-like domain, has also raised the possibility of a direct connection between Bcl-2 and proteases. BAG-1 has anti-cell death activity in transfection studies and cooperates with Bcl-2 in the suppression of apoptosis, providing enhanced protection from apoptotic stimuli beyond that conferred by either Bcl-2 or BAG-1 alone (Takayama *et al.*, 1995). One interesting revelation to come from studies of BAG-1 is that some cell death stimuli previously thought to function through a Bcl-2-independent mechanism, such as apoptosis induced by Fas or cytolytic T-cells, were inhibited by the combination of Bcl-2 and BAG-1 for protection. These findings suggest that in the absence of adequate levels of appropriate partner proteins, elevations in Bcl-2 protein levels may not be sufficient to render some types of cells resistant to some cell death stimuli. Other mechanisms such as high levels of Bax or Bak, production of dominant inhibitors of Bcl-2 (Bcl-X_S; BAD), or possibly post-translation modifications such as phosphorylation, could conceivably account for the failure of Bcl-2 to protect against apoptosis in some circumstances (Haldar *et al.*, 1995), thus begging the question of whether pathways for apoptotic cell death truly exist that do not involve Bcl-2 or one of its homologs at some level.

Finally, an additional potential mechanism of action for the Bcl-2 protein has been raised by the recent finding that Bcl-2 can physically associate with signal transducing proteins, including the GTPase R-Ras and the serine/threonine-kinase Raf-1 (Fernandez-Sarbia *et al.*, 1993; Wang *et al.*, 1994). In gene transfer experiments, a constitutively activated version of Raf-1 kinase was shown to synergize with Bcl-2 in preventing apoptosis induced by lymphokine withdrawal from a factor-dependent hemopoietic cell line, yet did not induce phosphorylation of the Bcl-2 protein (Wang *et al.*, 1994). Conversely, in the same cell, model, activated versions of R-Ras accelerated apoptosis through a mechanism that was completely suppressible by co-expression of Bcl-2 (Wang *et al.*, 1995). Interestingly, R-Ras failed to bind to either Bcl-2 or Bax in these cells, suggesting an indirect mechanism. These findings nevertheless suggest the possibility of a single transduction system, presumably centered around the membranes where Bcl-2 residues, including the outer mitochondrial membrane, nuclear envelope and ER, that may be uniquely involved in regulating cell death pathways, as opposed to the traditional roles for Raf-1 and other signalling proteins at the plasma membrane where they participate in signal transduction pathways linked to mitogenesis. Consistent with this idea are observations derived from the use of a cell-free system for apoptosis, where it was shown that mitochondria are required for apoptotic-like degradation of nuclei in cytosolic extracts prepared from *Xenopus* eggs (Newmeyer, 1994), implying that some kind of cell death "signal" was originating from the mitochondria. Experiments with respiratory chain inhibitors and free-radical scavengers suggested that reactive oxygen species were not significantly involved, whereas phosphotyrosine and Zn^{2+}, a known inhibitor of some protein tyrosine phosphatases, were effective at preventing apoptotic distruction of nuclei. Precisely how Bcl-2 might participate in the regulation of a hypothethical signal transduction system centered around mitochondrial and other intracellular membranes remains to be established. In this regard, it remains unproven that the interaction of Bcl-2 with Raf-1, or most of the other recently identified Bcl-2-interacting proteins such a Nip-1, Nip-2, and Nip-3, is essential for Bcl-2's function as an inhibitor of cell death. Only when the domains involved in these protein-protein interactions have been precisely mapped and appropriate mutagenesis studies performed, will the relative importance of these interactions be revealed. Information of this type represents an essential first step towards the ultimate goal of identifying novel pharmaceuticals that may one day improve our ability to treat cancer and many other diseases that involve dysregulation of the physiological cell death pathway.

Acknowledgements

We wish to thank C. Stephens for manuscript preparation, and the American Cancer Society, Ca-P CURE, Leukemia Society of America, and National Cancer Institute for generous support.

References

Baffy, G., Miyashita, T., Williamson, J.R., and Reed, J.C. 1993, Apoptosis induced by withdrawal of Interleukin-3 [IK-3] from an IL-3-dependent hematopoietic cell line associated with repartitioning of intracellular calcium is blocked by enforced Bcl-2 oncoprotein production *J. Biol. Chem.*, **268**:6511–6519.

Bodrug, S.E., Aimé-Sempé, C., Sato, T., Krajewski, S., Hanada, M., and Reed, J.C., 1995, Biochemical and functional comparisons of Mcl-1 and Bcl-2 proteins: evidence for a novel mechanism of regulating Bcl-2 family protein function, *Cell Death Differ.*, **2**:173–182.

Boise, L.H., Gonzalez-Garcia, M., Postema, C.E., Ding, L., Lindsten,T., Turka, L.A., Mao, X., Nunez, G., and Thompson, C.B., 1993, *bcl-2-* related gene that functions as a dominant regulator of apoptotic cell death, *Cell*, **74**:597–608.

Borner, C., Martinou, I., Mattmann, C., Irmler, M., Scharer, E., Martinou, J.-C., and Tschopp., J., 1994. The protein bcl-2α does not require membrane attachment, but two conserved domains to suppress apoptosis, *J. Cell. Biol.*, **126**:1059–1068.

Boyd, J.M., Malstrom, S., Subramanian, T., Venkatesh, L.K., Schaeper, U., Elangovan, B., D'Sa-Eipper, C., and Chinnadurai, G., 1994, Adenovirus E1B 19 kDa and Bcl-2 proteins interact with a common set of cellular proteins, *Cell*, **79**:341–351.

Cazals-Hatem, D., Louie, D., Tanaka, S., and Reed, J.C., 1992, Molecular cloning and DNA sequence analysis of cDNA encoding chicken homolog of the *bcl-2* oncoprotein, *Biochim. Biophys. Acta*, **1132**:109–113.

Chiou, S.-K., Tseng, C.-C., Rao, L., and White, E., 1994, Functional complementation of the adenovirus E1B 19-kilodalton protein with Bcl-2 in the inhibition of apoptosis in infected cells, *J. Virol.*, **68**:6553–6566.

Chittenden, T., Harrington, E.A., O'Connor, R., Flemington, C., Lutz, R.J, Evan, G.I., and Guild, B.C., 1995, Inductin of apoptosis by the Bcl-2 homologue Bak, *Nature*, **374**:733–736.

de Jong, D., Prins, F.A., Mason, D.Y., Reed, J.C., van Ommen, G.B., and Kluin, P.M., 1994, Subcellular localization of the bcl-2 protein in malignant and normal lymphoid cells, *Cancer Res.*, **54**:256–260.

Dowd, D.R., MacDonald, P.N., Komm, B.S., Haussler, M.R., and Miesfeld, R.L., 1992, Stable expression of the calbindin-D28K complementary DNA interfers with the apoptotic pathway in lymphocytes, *Mol. Endocrinol.*, **6**:1843–1848.

Farrow, S.N., White, J.H.M., Martinou, I., Raven, T., Pun, K.-T., Grinham, C.J., Martinou, J.-C., and Brown, R., 1995, Cloning of a bcl-2 homologue by interaction with adenovirus E1B 19K, *Nature*, **374**:731–733.

Fernandez-Sarbia, M.J., and Bischoff, J.R., 1993, Bcl-2 associates with the ras-related protein R-ras p23, *Nature*, **366**:274–275.

Gillet, G., Guerin, M., Trembleau, A., and Brun, G., 1995, A Bcl-2-related gene is activated in avian cells transformed by the Rous sarcoma virus, *EMBO J.*, **14**:1372–1381.

Haldar, S., Jena, N., and Croce, C.M., 1995, Inactivation of Bcl-2 by phosphorylation, *Proc. Natl. Acad. Sci. USA*, **92**:45087–4511.

Hanada, M., Krajewski, S., Tanaka, S., Cazals-Hatem, D., Spengler, B.A., Ross, R.A., Biedler, J.L., and Reed, J.C., 1993, Regulation of *bcl-2* oncoprotein levels with differentiation of human neuroblastoma cells. *Cancer Res*, **53**:4978–4986.

Hanada, M., Aimé-Sempé, C., Sato, T., and Reed, J.C., 1995, Structure-function analsyis of *bcl-2* protein: identification of conserved domains important for homodimerization with *bcl-2* and heterodimerization with *bax*, *J. Biol. Chem.*, **270**:11962–11968.

Hockenbery, D., Oltvai, Z., Yin, X.-M., Milliman, C., and Korsmeyer, S.J., 1993, Bcl-2 functions in an antioxidant pathway to prevent apoptosis, *Cell*, **75**:241–251.

Jacobson, M.D., and Raff, M.C., 1995, Programmed cell death and Bcl-2 protection in very low oxyten, *Nature*, **374**:814–816.

Jacobson, M.D., Burne, J.F., and Raff, M.C., 1994a, Programmed cell death and Bcl-2 protection in the absence of a nucleus, *EMBO J.*, **13**:1899–1910.

Kane, D.J., Sarafin, T.A., Auton, S., Hahn, H., Gralla, F.B., Valentin, J.C., Ord, T., and Bredesen, D.E., 1993, Bcl-2 inhibition of neural cell death:decreased generation of reactive oxygen species, *Science*, **262**:1274–1276.

Kiefer, M.C., Brauer, M.J., Powers, V.C., Wu, J.J., Ubansky, S.R., Tomei, L.D., and Barr, P.J., 1995, Modulation of apoptosis by the widely distributed Bcl-2 homologue Bak, *Nature*, **374**:736–739.

Kitada, S., Takayama, S., DeRiel, K., Tanaka, S., and Reed, J.C., 1994, Reversal of chemoresistance of lymphoma cells by antisense-mediated reduction of bcl-2 gene expression, *Antisense Res. Dev.*, **4**:71–79.

Kitada, S., Krajewski, S., Miyashita, T., Krajewska, M., and Reed, J.C., 1996, γ-radiation induces upregulation of Bax protein and apoptosis in radiosensitive cells *in vivo*, *Oncogene*, **12**:187–192.

Kozopas, K.M., Yang, T., Buchan, H.L., Zhou, P., and Craig, R., 1993, Mcl-1, a gene expressed in programmed myeloid cell differentiation, has sequence similarity to bcl-2, *Proc. Natl. Acad. Sci. USA*, **90**:3516–3520.

Krajewski, S., Tanaka, S., Takayama, S., Schibler, M.J., Fenton, W., and Reed, J.C., 1993, Investigations of the subcellular distribution of the bcl-2 oncoprotein: residence in the nuclear envelope, endoplasmic reticulum, and outer mitochondrial membranes, *Cancer Res.*, **53**:4701–4714.

Krajewski, S., Blomvqvist, C., Franssila, K., Krajewska, M., Wasenius, V-M., Niskanen, E., and Reed, J.C., 1995a. Reduced expression of pro-apoptotic gene Bax is associated with poor response rates to combination chemotherapy and shorter survival in women with metastatic breast adenocarcinoma, *Cancer Res.,*, **55**:4471–4478.

Krajewski, S., Mai, J.K., Krajewska, M., Sikorska, M., Mossakowski, M.J., and Reed, J.C., 1995b, Upregulation of bax protein levels in neurons following cerebral ischemia, *J. Neurosci.*, **15**:6364–6376.

Lam, M., Dubyak, G., Chen, L., Nunez, G., Miesfeld, R.L., and Distelhorst, C.W., 1994, Evidence that Bcl-2 represses apoptosis by regulating endoplasmic reticulum-associated Ca^{2+} fluxes, *Proc. Natl. Acad. Sci. USA*, **91**:6569–6573.

Lin, E.Y., Orlofsky, A., Berger, M.S., and Prystowsky, M.B., 1993a, Characterization of A1, a novel hemopoietic-specific early-response gene with sequence similarity to *bcl-2*, *J. Immunol.*, **151**:1979–1988.

Macleod, K.F., Sherry, N., Hannon, G., Beach, D., Tokino, T., Kinzler, K., Vogelstein, B., and Jacks, T., 1995, p53-dependent and independent expression of p21 during cell growth, differentiation, and DNA damage, *Genes and Dev.*, **9**:935–944.

Magnelli, L., Cinelli, M., Turchetti, A., and Chiarugi, V.P., 1994, Bcl-2 overexpression abolishes early calcium waving preceding apoptosis in NIH-3T3 murine fibroblasts, *Biochem. Biophys. Res. Commun.*, **204**:84–90.

Meikrantz, W., Gisselbrecht, S., Tam, S.W., and Schlegel, R., 1994, Activation of cyclin A-dependent protein kinases during apoptosis, *Proc. Natl. Acad. Sci. USA*, **91**:3754–3758.

Miura, M., Zhu, H., Rotello, R., Hartwieg, E.A., and Yuan, J., 1994, Induction of apoptosis in fibroblasts by IL-1 b-converting enzyme, a mammalian homolog of the *C. elegans* cell death gene *ced-3*, *Cell*, **75**:653–660.

Miyashita, T., and Reed, J.C., 1992, Bcl-2 gene transfer increases relative resistance of S491. and WEH17.2 lymphoid cells to cell death and DNA fragmentation induced by glucocorticoids and multiple chemotherapeutic drugs, *Cancer Res.*, **52**:5407–5411.

Miyashita, T., and Reed, J.C., 1993, Bcl-2 oncoprotein blocks chemotherapy-induced apoptosis in a human leukemia cell line, *Blood*, **81**:151–157.

Miyashita, T., and Reed, J.C., 1995, Tumor suppressor p53 is a direct transcriptional activator of human bax gene, *Cell*, **80**:293–299.

Miyashita, T., Krajewski, S., Krajewski, M., Wang, H.G., Lin, H.K., Hoffman, B., Lieberman, D., and Reed, J.C., 1994a, Tumor suppressor p53 is a regulator of *bcl-2* and *bax* in gene expression *in vitro* and *in vivo*, *Oncogene*, **9**:1799–1805.

Miyashita, T., Kitada, S., Krajewski, S., Horne, B., Delia, D., and Reed, J.C., 1995, Over expression of the Bcl-2 protein increases the half-life of p21[Bax]. *J. Biol. Chem.*, **270**:26049–26052.

Miyashita, T., Harigai, M., Hanada, M., and Reed, J.C., 1994b, Identification of a p53-dependent negative response element in the *bcl-2* gene, *Cancer Res.*, **54**:3131–3135.

Newmeyer, D., Farschon, D.M., and Reed, J.C., 1994, Cell-free apoptosis in Xenopus egg extracts by Bcl-2 inhibits a latent cytoplasmic phase, *Cell*, **79**:353–364.

Oltvai, Z., Milliman, C., and Korsmeyer, S.J., 1993, Bcl-2 heterodimerizes *in vivo* with a conserved homolog, Bax, that accelerates programmed cell death, *Cell*, **74**:609–619.

Reed, J.C., 1994, Bcl-2 and the regulation of programmed cell death, *J. Cell Biol.*, **124**:1–6.

Reed, J.C., 1995, Bcl-2: Prevention of apoptosis as a mechanism of drug resistance, in: *Hematology-Oncology Clinics of North America*, V. 9, G. Fisher and B. Siziki, ed. pp. 451–473.

Ryan, J.J., Prochownik, E., Gottlieb, C.A., Apel, I.J., Merino, R., Nunez, G., and Clarke, M.F., 1994, c-myc and bcl-2 modulates p53 function by altering p53 subcellular trafficking during the cell cycle, *Proc.Natl. Acad. Sci. USA*, **91**:5878–5882.

Sato, T., Hanada, M., Bodrug, S., Irie, S., Iwama, N., Boise, L.H., Thompson, C.B., Golemis, E., Fong, L., Wang, H.-G., and Reed, J.C., 1994, Interactions among members of the *bcl-2* protein family analyzed with a yeast two-hybrid system, *Proc. Natl. Acad. Sci. USA*, **91**:9238–9242.

Sato, T., Hanada, M., Bodrug, S., Irie, S., Iwama, N., Boise, L.H., Thompson, C.B., Golemis, E., Fong, L., Wang, H.-G., and Reed, J.C., 1995, Correction: Interactions among membes of the Bcl-2 protein family analyzed with a yeast two-hybrid system, *Proc. Natl. Acad. Sci. USA*, **92**:1793.

Selvakumaran, M., Lin, H.-K., Miyashita, T., Wang, H.-G., Krajewski, S., Reed, J.C., Hoffman, B., and Liebermann, D., 1994, Immediate early up-regulation of bax expression by p53 but not TGFb1:a paradigm for distinct apoptotic pathways, *Oncogene*, **9**:1791–1798.

Shimizu, S., Eguchi, Y., Kosaka, H., Kamiike, W., Matsuda, H., and Tsujimoto, Y., 1995, Prevention of hypoxia-induced cell death by Bcl-2 and BclxL, *Nature*, **374**:811-813.

Takayama, S., Cazals-Hatem, D.L., Kitada, S., Tanaka, S., Miyashita, T., Hovey, L.R., Huen, D., Rickinson, A., Veerapandian, P., Krajewski, S., Saito, K., and Reed, J.C., 1994, Evolutionary conservation of function among mammalian, avian, and viral homologs of the bcl-2 oncoprotein: structure-function implications, *DNA Cell Biol.*, **13**:679–692.

Takayama, S., Sato, T., Krajewski, S., Kochel, K., Irie, S., Millan, J., and Reed, J.C., 1995, Cloning and functional analysis of BAG-1: a novel Bcl-2 binding protein with anti-cell death activity, *Cell*, **80**:279–284.

Tanaka, S., Saito, K., and Reed, J.C., 1993, Structure-function analysis of the apoptosis-suppressing bcl-2 oncoprotein: Substitution of a heterologous transmembrane domain restores function to truncated Bcl-2 proteins, *J. Biol. Chem.*, **268**:10920–10926.

Tsujimoto, Y., and Croce, C.M., 1986, Analysis of the structure, transcripts, and protein products of bcl-2, the gene involved in human follicular lymphoma, *Proc. Natl. Acad. Sci. USA*, **83**:5214–5218.

Tsujimoto, Y., Louie, E., Bashir, M.M., and Croce, C.M., 1988, The reciprocal partners of both the t(14;18) and t(11;14) translocation involved in B-cell neoplasms are rearranged by the same mechanism, *Oncogene*, **2**:347–351.

Vaux, D.L., Cory, S., and Adams, J.M., 1988, Bcl-2 gene promotes haemopoietic cell survical and cooperates with c-myc to immortalize pre-B cells, *Nature*, **335**:440–442.

Wang, H.-G., Miyashita, T., Takayama, S., Sato, T., Torigoe, T., Krajewski, S., Tanaka, S., Hovey, III, L., Troppmair, J., Rapp, U.R., and Reed, J.C., 1994, Apoptosis regulation by interaction of *bcl-2* protein and Raf-1 kinase, *Oncogene*, **9**:2751–2756.

Wang, H.-G., Millan, J.A., Cox, A.D., Der, C.J., Rapp, U.R., Beck, T., Zha, H., and Reed, J.C., 1995, R-ras promotes apoptosis caused by growth factor deprivation via a Bcl-2 suppressible mechanism, *J. Cell Biol.*, **129**:1103–1114.

Yang, E., Xha, J., Jockel, J., Boise, L.H., Thompson, C.B., and Korsmeyer, S.J., 1995, Bad: a heterodimeric partner for Bcl-X$_L$ and Bcl-2, displaces *bax* and promotes cell death, *Cell*, **80**:285–291.

Yin, X.M., Oltvai, Z.N., and Korsmeyer, S.J., 1994, BH1 an dBH2 domains of *bcl-2* are required for inhibition of apoptosis and heterodimerization with *bax, Nature*, **369**:321-333.

Yuan, J., Shaham, S., Ledoux, S., Ellis, H.M., and Horvitz, H.R., 1993, The *C. elegans* cell death gene *ced-3* encodes a protein similar to mammalian interleukin-1 beta-converting enzyme, *Cell*, **75**:641–652.

Zhan, Q., Fan, S., Bae, I., Guillout, C., Liebermann, D.A., O'Connor, P.M., and Fornace Jr., A.J., 1994, Induction of *bax* bygenotoxic stress in human cells correlates with normal p53 status and apoptosis, *Oncogene*, **9**:3743–3751.

Index